The Wonder of the World

A Journey from Modern Science to the Mind of God

The **Wonder**
of the **World**

A Journey from Modern Science

to the Mind of God

ROY ABRAHAM VARGHESE

FOUNTAIN HILLS, ARIZONA

Tyr Publishing
PO Box 19895
Fountain Hills, AZ 85269-9895
Phone: 1 (800) Dial-A-Book (342-5226)
Fax: (480) 816-6187
http://www.tyrpublishing.com/
info@tyrpublishing.com

Ordering information
To order additional copies, contact your local bookstore or see our Web site. Quantity discounts are available.

Interactive Web site:
http://www.thewonderoftheworld.com/

ISBN: 0-9723473-1-3

Library of Congress Cataloging-in-Publication Data

Varghese, Roy Abraham.
   The wonder of the world : a journey from modern science to the mind of God
/ Roy Abraham Varghese.
      p. cm.
   Includes bibliographical references and index.
   ISBN 0-9723473-1-3 (hardcover : alk. paper)
   1. God--Proof, Cosmological.   I. Title.
   BT103.V37 2003
   201'.65--dc22

2003017887

Publication Date: 02/01/04

⊗   This paper meets the requirements of ANSI/NISO Z39.48-1992 (Permanence of Paper).

Book design by Xenocast
http://www.xenocast.com/

On August 28 and 29, 2000, New York City was the venue of the Millennium World Peace Summit of Religious and Spiritual Leaders hosted by the United Nations. Discourses and discussions on the role of religion in the world were the main events at this gathering. Professor Madhva Mitra, founder of the Sakshi Hermitage in the Himalayas and adjunct professor of physical science and natural theology at Wykeham College in Oregon, and Joseph Levin, a post-doctoral research assistant in Artificial Intelligence at the Massachusetts Institute of Technology, co-chaired a panel on the relationship of science and religion. Mitra held that modern science has culminated in the religious view of reality with which it began. In stark contrast, Levin argued that religious belief is a superstition discredited by science; furthermore, science has progressed in spite of religion not because of it. The dialogue soon reached a dead-end.

In November 2000, Levin lost his father Howard. Torn by the anguish of death's dateless night and an inability to believe in any kind of hope and meaning beyond the material universe, Levin resumed his dialogue with Mitra.

Mitra first outlined three exciting new discoveries:

1. Modern science emerged from a matrix articulated and precisely formulated by the four greatest thinkers of four major world religions, Hinduism, Judaism, Christianity and Islam. This matrix, the womb of science, may be thought of as a supra-scientific Theory of Everything that begins with the fundamental "God equation."

2. Science itself bears unmistakable witness to a fundamental truth embodied by this matrix: the universe was brought into being by an infinite Intelligence, God.

3. It is possible to see God here and now through the Wonder of the World.

In light of these discoveries, Mitra and Levin discuss recent findings on the origins of the universe, energy, life, mind and the laws of nature. The renewed dialogue is now a journey through modern science.

This is an account of the journey.

To the Father from Whom all things come

Table of Contents

୧୨

Prologue

ear Joe:

I cannot tell you how shocked and grieved I was to learn about your terrible loss. In one breathtaking, heartbreaking instant he left us, without warning or farewell. In one mind-numbing moment he vanished from our midst. In a trance, we lowered his beloved body into a bottomless pit. In total wonderment, we heard friends and family discuss him in the past tense. In a haze of horror, our thoughts came to grips with our feelings: we will never see him again in this world, his smile will never cheer our spirits one last time, his tenderness will never again touch our souls.

I'm aware that nothing I say can begin to console you in the face of the void left by his tragic and untimely departure. But if there can be no consolation for immediate grief, I believe there's a glimmer of light that gently lets itself in through the darkness if our minds are open. I'm deeply gratified that you're willing to let me share the basis of this belief. And, of course, the sharing will be mutual because we can't reach a common understanding without dialogue. Let me be very frank at the outset by saying that it's my hope and prayer that the loss of your beloved father will be followed by your discovery of the Father of all life and being.

At a fundamental level, death is the dark shadow hanging over every human thought, choice, plan and action, seeming to rob them of ultimate meaning and value. I know that you believe we humans have come from nowhere and from nothing and are likewise destined to return nowhere, become nothing; that all we have achieved and become will be lost forever when we die so that it will be as if we had never been.

In contrast, and in common with all the major religions, I hold that human life does have an ultimate purpose, meaning and significance, and that the true "end" of Homo sapiens, the raison d'être of human existence, is union with the Ultimate Reality. We do not come from nowhere and nothing and return nowhere to nothing. We come from God and our fulfillment comes in a union with God, a union that preserves our individual identity forever. As the theist of any religion sees it, true fulfillment for the human person comes from a union of love with the infinite plenitude of perfections who is both Beginning and End.

Theism, the belief in Theos or God, an infinite, omnipotent, omniscient, omnipresent, personal Being, is not to be found only in religions of the Book, Judaism, Christianity and Islam. Some of the most eloquent expositions of theism have come to us from the great Hindu thinker Madhvacharya. It is, in fact, a gross mistake to identify Hinduism exclusively with monism or pantheism since five out of six of the Hindu schools affirm that our souls cannot be identified with the Ultimate Reality. Hinduism's *Rig Veda* (1200-900 B.C.), the first major work in an Indo-European language, speaks of "our father, who created and set in order and knows all forms, all worlds." The same thought is to be found in the texts of the ancient Hebrews: "Is not this your father, who gave you being, who made you, by whom you subsist?" (Deuteronomy 32:6). China too had its own noble school of theism, Moism, dating back to the fifth century b.c. And studies of the most primeval religions across the world indicate that, in many instances, they were driven by belief in a Supreme God commonly addressed as Creator and Father.

I propose to start our journey with a presentation of the Seven Wonders, the wonders embedded in our experience and everywhere present that constitute what I call the Wonder of the World. As I've often said, modern science is the strongest witness to this vision of the wonder of the world, a vision that culminates in the recognition of an infinite Intelligence that invents, unifies and empowers all things. I hope to show also that this mysterious and fascinating Reality, the "mysterium tremendum et fascinans," as Rudolf Otto put it, can be directly encountered in our experience at all levels. That is to say, we can "see" God here and now, although this "seeing" is inevitably mediated through the structures and processes of the world.

If "Seven Wonders" narrates what we experience in daily life and learn from science, the next section, "Sages and Scientists," recounts humanity's response to the Wonder of the World. I focus specifically on the way in which this response inspired and shaped modern science. My thesis is that the foundational framework of modern science, with the key idea of laws of nature, was born and bred in the theistic world-vision. What is more, prior to this and within a time window of 300 years, the four finest thinkers of Hinduism, Judaism, Christianity and Islam framed a meta-scientific Theory of Everything that underpins the scientific enterprise. This intellectual superstructure, which we shall call the Matrix, provided a systematic rationale for the foundations of science. Its starting-point and core principle was an "equation of God." Interestingly the great scientists who founded modern science, Copernicus, Newton, Maxwell,

Einstein, Planck, Heisenberg, Dirac and numerous others, were Prophets of the Matrix in the sense that they passionately proclaimed the root-and-fruit embeddedness of science and religion. *The Matrix is the common platform that supports both science and religion.* To be sure the Matrix has had one major foe, Monism. *Monism is the idea that the world is made strictly of one kind of substance.* Monism comes in two varieties. The spiritualist monist says we are all part of one spirit. The materialist monist says we are all temporary manifestations of matter. The spiritualist says that matter is a delusion; the materialist that mind is an illusion. Monism is the great undead philosophy of our time, not living but not quite dead. And both varieties of monism are fatal for science.

So there you have it, the first half of my agenda. With the conclusion of my initial presentations, as discussed, we can commence our interactive study of the world revealed by modern science.

I will present my understanding of the following:

- ♦ The phenomenon of sight
- ♦ Space, time, motion and matter
- ♦ The quantum microcosmos
- ♦ Modern cosmological accounts of the origin of the universe
- ♦ The origin of life, consciousness and mind
- ♦ The laws of nature
- ♦ The existence of an infinite Intelligence.

I will be concerned with the data available to us and what I think are valid interpretations of the data.

Through all this, I hope to engage you in dialogue so as to better understand what you consider to be obstacles. I firmly believe that most disagreements in this area stem from semantics and misconceptions. Since you've kindly agreed to keep an open mind to the point of even being sympathetic, I'm confident we will make progress. A debate of adversaries and antagonists, on the other hand, is less a journey to a destination than it is a workout on a treadmill.

Since you've been too gracious to mention it, let me put the issue of my credentials on the table. As you well know, I'm not a professional scientist. So why should you have any confidence in my handling of scientific concepts and theories? Now I certainly think that generalists should beware of trespassing on specialist turf. But I have no intention of entering into or pronouncing on disputes within any scientific specialty. I freely admit that I'm not competent to do so.

But I certainly am competent to make judgments concerning logical or ontological interpretations of the data from science. Equally, any extrapolations and speculations that can't be verified on the basis of available data will be subject to critical scrutiny, especially when these involve meta-scientific issues. By ontological, I mean those essential and ultimate principles that underlie science but cannot be proved by science; thus science assumes but cannot prove that the world exists since any proof implicitly assumes the world's existence. By meta-scientific, I mean anything that cannot be tested with scientific tools or methods; it cannot even be understood in quantitative terms. Science can tell us about the physical history of the universe. But the question of why the universe exists is a meta-scientific question and these kinds of questions are addressed by ontology. Einstein, the greatest scientist of the last 100 years, was also one of the most honest; he confessed quite bluntly in his *Out of My Later Years* that the man of science is a poor philosopher. Here I accept the hard data provided by the professional scientists. My comments focus on the interpretation of such data, on the models for understanding and explaining the data from science.

A few words about our communication mode. In view of the volumes of data involved, I agree with you that an online chat model won't work. So your idea of setting up a private newsgroup on your school server with a connection to the Internet seems the most viable solution. If I understood you correctly, this allows us to send or read via email or via a web interface according to our preference. As you know, large parts of my presentations were already prepared for this manuscript I've been working on since I was born(!). I'll send my presentations in the sequence I've outlined and, once you've downloaded them, you can send me your responses. If it's OK with you, we'll start the interactive discussions once we get to my "Seeing is Believing" presentation.

I'm amused (and flattered) by your suggestion of screen names for our online discussions, "Guru" for myself and "Geek" for you! Well, I have no right to such a title, but I'll gladly accept it in the same humorous spirit in which you proposed it.

In union of truth and friendship,

Madhva

SEVEN WONDERS

*T*o wonder is to know and to know is to wonder. To wonder at the world is to know its deepest secrets for, as Albert Einstein once said, the sense of wonder in the face of the mysterious is the cradle of true art and true science.

But the modern world knows little of wonder. Some grinch has stolen the magic that makes us wonder and turned the paradise we call the world into a desolate wilderness. My mission here and now is to restore this greatest of gifts to its rightful owners, to re-enchant the world.

But first we must identify the culprit who has made off with the treasure. We are told by the initiated that the culprit is something called modern science. And here let me make the first of my bold assertions. In my view, the suspect, modern science, is not just innocent until proved guilty but is, in fact, our ablest ally in recovering our stolen inheritance. It's not science that is the grinch but a band of intellectuals trapped in vacuous abstractions and irrational ideologies.

For the world revealed to us by science is a world of wonder and glory, magic and mystery. And this is the world that must be recovered from the hands of those who've blinded us and buried the evidence.

Let me illustrate my thesis very simply. Take a look outside your window. What science tells you is that the field of grass you see is a bank of sophisticated computers processing vast quantities of information and, unlike computers, responding intelligently and instantly to a whole host of sensory inputs from gravity to sunshine to an array of chemicals. The sky is blue because of a vast network of fundamental constants in nature; the sun shines because it turns millions of kilograms of mass into energy every second, and every cubic centimeter of the seemingly empty space that serves as a backdrop is teeming with energy fields.

At the foundations and frontiers of modern science we find five great mysteries: energy, life, consciousness, mind and the intelligence embodied in the laws of nature. Science describes and documents the marvelous reality of these five and of other everyday mysteries like language, reproduction and seeing. But science cannot tell us how any of them emerged or what is their source or, in the case of the first five, even define their nature. Once we truly recognize their reality, however, we realize that each one of these fundamental hard facts can be explained only in terms of Something that is Itself Intelligent, Powerful, Living, Conscious and Rational.

This insight has not been lost on many of the greatest scientists of modern times, from Einstein to Stephen Hawking, who have identified the laws of nature with the Mind of God. The journey from the world of modern science to the Mind of God is the theme of both my presentation of the Seven Wonders and our dialogue. Science is an essential companion on our journey because the wonder and glory of the world can be restored only when we come to see the stupendous Intelligence that underlies it at all levels.

To put it simply, I hold that infinite Intelligence is at work in the world manifesting itself in multitudinous ways, and that this is evident both in our common experience and the data of the sciences. In fact, the wonder of the world is a window through which we can and do see God. This final contention is the climax of my presentation of the seven wonders.

Chapter 1

How I Wonder That You Are

"NOT HOW THE WORLD IS but that it is is the mystical," said the great philosopher Ludwig Wittgenstein. This is our starting-point, one that's obvious if you think about it, but rarely thought about. When you read the newspapers or the best-selling books of our time there's no discussion at all about the most obvious and the most fundamental mystery of all: that we exist!

Now wonder is neither a poetic nor a mystical phenomenon. It's as natural and necessary for human beings as the capacity to think and to feel. To lose the capacity of wonder is, in fact, to lose our greatest avenue of knowledge, for the most important truths can be known only if we have a minimal comprehension of their grandeur. Thus, the realization that anything exists at all is an apprehension wrapped in the most incredible aura of mystery.

Let's say that the universe was always here. But *how* could it thus be here and be here in its present form, no matter how it came to take its present form? If it had no beginning, we still want to know how an entity with such properties came to exist. The normal mind cannot be satisfied with the idea of the universe simply existing as a brute fact, especially a universe of beings with consciousness and intelligence. Only brutes take refuge in brute facts. The fact that anything at all exists points ineluctably to a need to explain the existence of this anything. Such an explanation can ultimately only be found in the existence of a Being that explains both its own existence and that of everything else, a Being who exists necessarily and has all the perfections of existence.

But then we wonder how it came to be that there is such a Being. Conceptually we can come to see that there could be such a necessarily existing Being that explains the existence of anything at all. But to know *that* is not to know *how*. To think about this Being whom we call God is to be struck by wonder in the face of overwhelming Mystery, the truth that God always existed and the equally bewildering fact that all of time and the things of time were seen by God from all eternity.

The wonder of existence is a mystery that confronts us at every instant. Every time we realize that we and anything around us exist, we are moved to wonder how this was and is possible and how it all began. We know the only possible explanation is that there has to be an Ultimate Reality that always IS. The mystery of Its existence is infinitely more mysterious than our own existence; we know, however, that Its existence has to be the case if anything is to be the case. It makes sense of all things because it is Sense. Thus the recognition of the Infinite is not a leap in the dark but a leap to the light.

Chapter

How Does the Electron Know What to Do?

OW DOES THE ELECTRON know what to do? Why does each electron continue to maintain a specific charge shared with all other electrons and equal to that of the 1,836 times bigger opposite-charged proton? What makes it orbit a nucleus in a certain fashion and emit energy when it takes a quantum leap from one orbit to another? What tells two electrons to shoot out a photon when they run into each other? What enables each electron to enjoy a natural life span of ten billion trillion years?

At first glance we might be inclined to respond that electrons like everything else in the world simply follow the laws of nature, and no doubt we would be right at a certain level. But a moment's thought should tell us how utterly extraordinary it is that the physical constituents of this world obey any laws at all. Why on earth should stones and feathers and stars follow uniform laws of motion? What tells the molecules of a gas that the product of their volume and pressure should be proportional to their absolute temperature? Why does water have, and continue to have, all the properties, such as cohesiveness, temperature storage, solvency, low viscosity, chemical reactivity, that are essential for it to serve as the cradle of life?

What compels the hydrosphere to keep, through interconnected cycles and over four billion years, as chronicled by Michael Denton, a constant level of the 25 elements required for life? How do nucleic acids like DNA and proteins know that they can and should communicate and construct, repair and replicate? How do genes know when to switch on, and why do they continue to do so at just the right time? What compels organisms to adapt to environment? Why should these and all the other particles and forces, galaxies and gas clouds, cells and chemicals that make up our world follow instructions? Do they know they're supposed to act in that way? Were they programmed to obey orders, march in step? Trillions of quarks and cells, billions of galaxies, all of them doing what they're told to. But by whom and how?

Do the laws exist independently of the fields that manifest themselves as particles and forces? If so how do these entities know that they should follow the laws? Or are the laws programmed in them? But where do the programs reside? No matter how much we open up the fields in question we can't find any programs in them. How do gluons know that they should be carriers of the strong nuclear force that bind quarks? Do they see the writing on the wall? Which wall?

Let's grant that there are laws of nature. We might verbally deny them, but we certainly won't take the extraordinary step of proving our point by jumping off the fifth floor. But if there are laws, we confront the puzzling question of how it's possible for a material thing to follow any law whatsoever. We humans follow laws that we set up because we're conscious agents capable of intentional action. But how can an inanimate thing be made to do anything, let alone follow a law?

We might say that the laws are simply invariant properties of the universe and its constituents. But this semantic switch doesn't solve anything, because we still ask how these properties were instituted in the first place and, again, how all the entities in the world can be made to behave in accordance with any property?

A determined reductionist might say that the electron's properties were determined in the furnace of the Big Bang, or that the structure of the strong nuclear force was determined when there was a symmetry-breaking of the original superforce. So there's no great mystery about why these things act the way they do. But this response simply shows we haven't understood the point at issue. Let's take things back to the Big Bang or even before. You might say this original Event determined all future properties and laws. But then we ask how it's possible for the Event or its precursors to not simply institute universe-wide laws and properties but also enforce them on a permanent basis. For instance, how can mass-energy be made to conform to the law of its conservation?

If this conformity doesn't strike you as extraordinary, perhaps an illustration will help. Imagine if you will that you are taking a stroll on the beach. There is no breeze. The tide is out. Suddenly, to your amazement, all the grains of sand on the beach rise up and start assembling themselves into a series of precisely patterned buildings with symmetrical rooms and lighting mechanisms. Now your instinctive reaction to this phenomenon would be to try to explain it. What strikes you as astonishing is not simply the coming to be of a brilliantly architected set of buildings but the mystery of how mere grains of sand could pull it all off acting in tandem and following specific pathways. You conclude that the sand has somehow been programmed to act as it does. But then you ask, what gave birth to the program, how was it inserted at the granular level and who watches over the enterprise to ensure that there are no deviations from the program? You are stunned both by the existence of the program and the fact that it could be orchestrated and implemented at such a fundamental level.

This beach-house tale is fanciful. But its main failing is inadequacy, not exaggeration. A grain of sand, which has a size of about 100 microns, for instance, is made up of a million atoms and each atom is about a millionth of a millimeter. Each atom is a universe in its own right with protons and neutrons, themselves made up of quarks, and electrons that are all held together by electromagnetic forces. Atoms contain huge reservoirs of energy. The energy in a single gram of any kind of matter, reports the noted scientist Gerald Schroeder, can boil 34 billion grams of water into steam. It was not for nothing that the poet William Blake talked about seeing a world in a grain of sand. Which brings us back to the same questions we asked with respect to the analogy: How was the sand programmed to be the way it is and how was the program inserted and then enforced?

Miracles are spoken of as exceptions to the laws of nature, but it seems obvious that the laws of nature themselves constitute the greatest miracle of all. When the astronaut orbiting the moon reported that Isaac Newton was doing most of the driving, he was saying that the laws described by Newton were still in effect. But we have no idea how this could be the case, and we don't know where the laws came from.

The existence of the laws of nature is the single greatest mystery uncovered by science, one that baffles the most skeptical scientist as much as it does anyone else. Those scientists who have reflected most fully on the laws of nature, from Newton to Einstein to Hawking, have come to the obvious and inescapable conclusion that the laws of nature were constituted by an infinite Intelligence, the Mind of God. Said Hawking in an interview, "The overwhelming impression is one of order. The more we discover about the universe, the more we find that it is governed by rational laws." In his *A Brief History of Time,* he wonders what breathes fire into the equations of science and gives a universe for them to describe. The answer to the question of why the universe exists, he said, would reveal to us "the mind of God."   And Einstein declared that anyone seriously engaged in the pursuit of science becomes convinced that the laws of nature manifest the existence of a spirit vastly superior to that of human persons. God, said Richard Swinburne, is not invoked to explain the facts explained by science but to explain the fact that science **can** explain.

Chapter

Smart Universe

*M*Y THESIS is that the fundamental category underlying the Universe is intelligence. We search anxiously for signals from across the Universe that will show the presence of intelligent agents out there. However, the Universe itself with its laws and fields and forces is the ultimate Signal of Intelligence. At all levels, we witness a magnificent progression in degrees and kinds of intelligence that originate and culminate in an infinite Intelligence. The four basic kinds of intelligence we encounter in our experience are embedded, active, self-aware, and, undergirding all the others, infinite Intelligence.

Four Kinds of Intelligence

One of the great contributions of modern science to our understanding of the world is the revelation that intelligence is all pervasive. From the forces and particles of the subatomic realm that operate within a framework of precise symmetries to the inexhaustibly resourceful DNA that builds the living world, we live in a "smart" universe. If you ask what is meant by "intelligent," I will be the first to admit that the idea of intelligence has been extended and adapted in our present context. Coupled with the advancements in the scientific story of the world, technology has given us a new perspective on the idea of intelligent systems. These technological developments furnish a useful model with which to consider the intelligence inbuilt in Nature.

My point is simply this: the universe we live in is "smart." I use "smart" in the same sense in which we talk of "smart" cards, "smart" tags, "smart" cars and intelligent terminals. Of course, I have no intention here of implying that the universe is an entity over and above all the things that constitute it. Nor am I saying that the universe is alive or has a self.

Many modern inventions are called "smart" or "intelligent." For instance:

- An "intelligent" terminal processes data and inputs locally while a dumb terminal only transmits inputs and provides displays.

- A "smart" card contains an integrated microprocessor within the card for storing and processing information, which is transmitted at transaction points through a card reader or an antenna coil.

- A "smart" tag or label contains information on the contents of a package. It wirelessly communicates this data to checkout points or appliances, e.g. giving heating instructions to the microwave.

- An AI (Artificial Intelligence) system belongs to a class of hardware and software that emulates such instances of human reasoning and learning as problem-solving, game-playing, pattern recognition and natural language processing, i.e. our normal speech patterns.

Often enough, we apply analogies from the natural world in talking about the things we make. Computers are compared to brains, satellites to eyes. Now it's time to return the favor by comparing things in nature to today's technologies.

If we compare the characteristics of tags and machines that are called "intelligent" with the things that make up the natural world, we find that *nature is a network of intelligent systems.* The dictionary definition of intelligence as a mental power of learning and understanding can be applied at least analogously to both the artificial and the natural. A computer with memory and processing power learns by storing data and understands by processing the data. A living system that generates energy, repairs itself and reproduces has its own kind of memory and processing power. The billions of galaxies that behave in accordance with the laws of cosmology and particle physics are driven by the information contained in the fields that constitute them.

Admittedly the comparison to human intelligence is one of analogy, not identity, and some distinctions must be drawn if we wish to be coherent. The human person is a self that is conscious both of itself and its environment, that conceptualizes, understands and communicates abstract ideas, and that freely intends and executes good and bad actions. The same cannot be said of microprocessors, cells and galaxies. Nevertheless, the behavior of each one of the three is clearly different from an entirely inert system or empty space. What we need is some way to clearly and accurately classify these different phenomena.

In the scheme we propose here:

- **Embedded intelligence** describes *intelligent systems like atoms that process information by following certain laws, but are not autonomous.*

- Living beings in general can be described as having **active intelligence** because *they are independent agents that can maintain and replicate themselves and act in the world and "learn" from it.* We will also call them **autonomous agents**.

- Human beings may be thought of as exhibiting **self-aware intelligence** *not only because they are capable of abstract thought and free actions but also because they are self-conscious.*

- *The source of all the intelligence* in the world is **infinite Intelligence** or God.

At the very heart of intelligence is the ability to create, process and synthesize information. Conversely, all information has to originate from intelligence of some kind, and the source of the information system that is the universe is supreme Intelligence.

While active, self-aware and infinite intelligence are associated with living agents, embedded intelligence is entirely in-built and to that extent passive. We will consider embedded intelligence in more detail here, and active, self-aware and infinite intelligence in succeeding chapters.

❧

Embedded Intelligence

Embedded intelligence has the following characteristics:

1. Information storage in encoded form

2. Information processing including data retrieval

3. Information flows that generate events and active structures, that proceed from initial states to goal states on precise pathways

4. Synchronization of this information with other systems, collaboration

This description is not comprehensive but it is a plausible description of those systems we call "intelligent" in the real and virtual worlds. Although these systems are not the same as life or human intelligence, it is obvious that a system with these characteristics is information-rich, rationally structured and process-driven. And this we call embedded intelligence.

So how does embedded intelligence manifest itself? Through processes, structures and laws. At every level, the physical world is a world of processes and structures driven by foundational laws. This rationality is fundamental and all encompassing, not limited simply to peripheral phenomena but to all physicality. From atoms and cells to galaxies and ecosystems, we see intricately ordered activities and operations manifesting carefully defined laws; that is, the laws that affect a quark affect a galaxy. These laws are not simply brute facts but constitute a platform of parameters driven toward life. We are struck not simply by the existence of rational principles underlying the physical world but also by the fact that they operate within certain specific ranges to enable the existence of conscious, rational life. The world is rationally ordered, and moreover it is ordered towards the production of rational life. The laws themselves, it may be said, are intelligent and purposive. And we are not speaking of abstractions here but of realities that manifest their presence to us all the time in our immediate experience.

Processes, Structures, Laws

A few examples from the inanimate and the animate domains should help illustrate what we're talking about. Let's start with processes. Process is the central reality we observe and experience in operation all the time and everywhere. Processes occur at different levels: physical, chemical, and biological. Take these instances:

♦ At any given moment, solar systems and galaxies spin across the universe. There is a perfect, and inexplicable, balance between the outward centrifugal force and the inward force of these massive structures. This balance prevents them from colliding with each other or being thrown apart.

♦ As predicted by James Clerk Maxwell, the universe is awash with electromagnetic radiation. The incredibly varied kinds of radiation out there in nature follow the same laws that we exploit in applications ranging from radar to mobile telephony.

♦ In the operation of the five senses, seeing, hearing, smelling, tasting and touching, the fundamental event involved is the transformation of a mechanical stimulus into a nerve signal sent to the brain that is then converted into a conscious state. Despite decades of scientific study and progress made in understanding the network of proteins, ions, signals and cellular structures involved, the bridge between these two worlds, the external stimulus and the corresponding sense-perception, remains as much a mystery today as ever.

Just as immediate and obvious as processes are the structures that constitute the world. Everything we experience, from the microscopic to the macroscopic, is part of a structure.

♦ On the large-scale level, the universe comprises stars that group together in galaxies that in turn form clusters. According to the astrophysicist Robert Jastrow, the observable universe is made up of a staggering 100 to 200 billion galaxies. On the average, these galaxies are themselves made up of an average of 100 to 200 billion stars. Most galaxies are either spiral or elliptical in shape. As just noted, they are held together by the gravitational attraction of their stars, and their rotational motion prevents them from collapsing in on themselves.

♦ The brain has 100 billion neurons that use dendrons and axons to send information across one quadrillion synapses. The 100 billion neurons do 100 million MIPS, i.e., millions of instructions per second.

♦ Water, a molecule made up of two atoms of hydrogen and one of oxygen, is the miraculous solvent that serves as the medium of life. Water makes up 80 percent of the mass of our cells. Water is also believed to have played a key role in the formation of stars and solar systems.

Precise but simple laws that can be formalized in mathematical terms govern all physical processes and structures. Consider these examples:

♦ Maxwell held that all electromagnetic waves are produced by oscillating electric charges, with the difference between two waves determined by their frequency of oscillation. He concluded that light was also an electromagnetic wave, and created a set of equations that apply to electricity, magnetism and optics.

♦ The equivalence of matter and energy is one upshot of the Special Theory of Relativity. This relationship is expressed in the equation $e=mc^2$ which tells us that the energy contained in any material object at rest is derived by multiplying its mass m by c^2 the square of the speed of light, 186,283 miles per second.

♦ Einstein's General Theory of Relativity, with its conception of gravity as geometry, proposed that matter curved space, and then inferred both the total mass of the universe and the resultant radius of curvature from the density of matter and the gravitational constant.

♦ Quantum mechanics is a mathematical framework generated to describe the subatomic world that has been successfully deployed in predicting the existence, behavior and relationships of subatomic entities that behave as waves or particles depending on the measuring device. Quantum behavior has been mathematically formalized in three modes that are formally equivalent: Schrödinger's wave mechanics, Heisenberg's matrix mechanics, and Dirac's non-commutative algebra. Quantum physics reveals a world that is intelligible, although its intelligibility is of a different nature than the macroverse. The principle that all phenomena have an explanation is no less applicable at the micro than the macro level.

Nothing manifests the existence of embedded intelligence more clearly than the correspondence between abstract mathematical theories and real-world applications. All the laws of physics are mathematical, and yet they are effective at every level of the physical world. Einstein's General Theory of Relativity was a work of abstract mathematics; nevertheless, it was precisely confirmed in experiments years after it was first proposed. Scientific laws are a collection of symbols endowed with specific meanings that are used to describe and understand the world, and these laws are notable for their simplicity and symmetry. In fact, the laws of physics can work only if there is underlying symmetry. Symmetry is, interestingly, a dominant theme in the work of string theorists searching for a theory of everything. Moreover many scientists today see scientific laws as algorithms and physical systems as computational systems. According to the quantum physicist David Deutsch, "The universality of computation is the most profound thing in the universe."

These connections lead scientists to talk of a cosmic code and a cosmic blueprint. Thus Einstein remarked that those who have made advances in science are struck by a reverence for the rationality that manifests itself in existence. Eugene Wigner spoke of the unreasonable effectiveness of mathematics in the natural sciences. The mathematician Roger Penrose dismisses the idea that the success of mathematics can be attributed to its survival value, and attributes the accord between mathematics and physics to the underlying rationality of the world.

Information and Intelligence

For many scientists, the codes and blueprints latent in the workings of the world are best described as information. All the information on a particle, for instance, is represented by the wavefunction that describes its behavior. The quantum physicist John Archibald Wheeler has said that his lifetime in physics was divided into three periods: (1) "everything is particles," (2) "everything is fields," and, now, (3) "everything is information." Gerald Schroeder observes that Wheeler "sees the world as the 'it' (the tangible item) that came from a 'bit' (eight of which comprise a byte of information)." Information *precedes* its manifestation in matter. According to Wheeler, "every *it*—every particle, every field of force, even the space-time continuum itself—derives its function, its meaning, its very existence from binary choices, *bits*. What we call reality arises in the last analysis from the posing of yes/no questions." As Schroeder puts it, "The tree and every other part of nature express in physical form the wave-like ethereal energy from which they are fashioned. And that elemental energy is not other than the manifestation of the wisdom from which it is built."

Unmistakably, matter, mass/energy in this context, is the primary vehicle of information in the world. Whether it is information programmed by us, e.g. software, movies, books, or communicated by mysterious instruction manuals like DNA, or simply inbuilt, as with anything that follows the laws of nature, everything in the universe is controlled by coded information. But matter is purely a vehicle. How did it become a vehicle for codes and blueprints? We know it takes intelligence to decode the information transmitted by matter. But if decoding requires intelligence, how about the encoding? If information exists prior to matter, what is its source?

At one level, all matter can be thought of as particles. At a more fundamental level, we realize particles are manifestations of fields. Beyond that, we now know that it's all about information. The next breakthrough is realizing that the foundation of it all is intelligence. Implicit in all its phases of discovery is the greatest insight of modern science: *everything is intelligence.*

The phenomenon of sound is as good an illustration as any of inbuilt intelligence.

The warbling of birds, the chirping of crickets, the rustling of leaves, melodies that move the soul, shrieks of joy, the cell phone that "throws" our voice around the globe. How easily we take these for granted, how normal they seem – but what a difference they make to the way we think and feel and act. Imagine a world without sound, a world where nothing rings out or echoes or whispers or gurgles or rises to a crescendo. Imagine a world where we cannot enjoy the elegant cadences of Homer's *Iliad* or the majestic rhythms of a Mozart and you will see then how fundamental a role sound plays in human life and experience.

The source of all this "sound and fury" is simply a form of mechanical energy moving in waves through various media, including solids, liquids and gases. The vibration of something physical like the vocal cords produces sound. It is propagated by the displacement of particles like air molecules. The rate of vibration of the particles is the frequency of a sound wave. We hear when:

1. The outer ear "picks up" sound waves,

2. passes them onto the middle ear,

3. the middle ear turns them into internal vibrations and waves and transmits these to the inner ear,

4. in the inner ear, the cochlea transforms the waves into nerve impulses that are sent to the brain.

But how can a mere wave of mechanical energy become a package of meaning (language) or a thing of beauty (music)? How can a vibration in the air, a fluctuation of pressure serve as the foundation for a haunting melody that opens windows into another world? Every link in the complex chain of energies, particles and organs that produces and propagates sound is indispensable if we are to have the song of a nightingale, a Shakespearean sonnet or a sonata by Haydn. If the goal is the creation of sound, then a most improbable collection of entities has been assembled to meet it. Which leads us to ask: is our medium of communication itself a communication? Is the packaging of intelligence itself intelligently packaged?

So Where Did It Come From?

Now imagine this scenario: we assemble the world's most brilliant scientists, living and dead, Einstein and Edison, Maxwell and Marconi, Newton and von Neumann. Is it conceivable that, given any amount of time, they would be able to create a universe like ours with all its precise parameters and laws? First, they would have to generate a universe-building recipe, then find the ingredients required and, finally, cook the dish. No one seriously believes that it's even remotely possible for the best and the brightest to pull off this job. There's also the added complication that they would have to be in existence before they can create everything, themselves included. If we can't imagine the greatest human intellects bringing the universe with its incredible intelligence into being, why would we assign the role to a vacuum (but who brought that into being?) or chance (which is the absence of all intelligence)?

Chapter

The Enigma of Energy

ENERGY! It's everywhere, all the time, shooting out as particles and forces, galaxies and vacuums, plants and animals; powering big bangs and the periodic table, brains and gene pools, houses and cars; creating light and heat, motion and mass. For all this, we are barely aware of its presence. Still less do we know what it is and where it comes from.

Of course, we know its manifestations and cycles and the laws governing its behavior. We know that there is an electromagnetic force that holds our bodies together, a gravitational force that holds us to the earth, and a nuclear force that keeps atoms together. We know that things have stored energy, such as the kinetic energy of a body in motion. We know that different forms of energy are converted into each other given a conversion mechanism. We are familiar with such displays of energy as velocity and temperature and pressure. We also know the following:

1. The formation of helium nuclei in the sun creates heat energy.

2. This heat is emitted as photons, a form of electromagnetic energy, to our planet.

3. On earth, through photosynthesis, the photons help create chemical energy in plants.

4. This passes on to animals when they consume plants to perform energy-expending activities.

Energy from the sun stored as chemical energy in coal or oil is converted back into heat, and, finally, into electricity in a power plant. We know also that the sum total of mass and energy in any system will remain the same.

With his $e = mc^2$ equation, Einstein showed us that matter is concentrated energy. It can be converted into other forms of energy, just as these other forms can be transformed into matter. The mass of a particle is a function of the amount of energy required to create it. As mentioned before, we see this interchangeability at work every day when we look at the sun; every second, it turns 44 million kilograms of mass into energy. But some questions remain. Although Newton and Einstein formulated the laws governing the motion and interaction of matter, we do not know why either energy or matter is subject to inertia and gravity. Why do they resist motion or feel gravitational force?

More important, what is energy? Every definition ends up being circular or redundant. If you say it is the capacity for doing work, it turns out that work is defined in terms of force and force in terms of energy. This is because energy is a fundamental concept like life. We can describe the properties associated with it like velocity and temperature. But we can't define it in terms of something more basic than itself.

And where did it come from? We're told that it all started with the Big Bang. But what powered the Big Bang? One theory has it that a fluctuation in a quantum vacuum triggered off the Big Bang event; in certain variations, the vacuum plays a role only after the Big Bang. But this theory supposes also that the vacuum itself is seething with energy fields, so the question becomes: how did the energy fields originate? We know not whence or how. The renowned astronomer Martin Rees acknowledges that it is basically mysterious how empty space could have energy associated with it. The universe is indeed composed of fields and energy, says the biochemist Rupert Sheldrake, but we do not know what is the primal nature and source of the energy or the origin of the primal field. This, then, is the enigma of energy.

Here I will present three perspectives on energy:

1. an overview of its primary vehicle, fields;

2. its most fascinating manifestation, light; and

3. its embodiment as motion.

Fields

The Standard Model of modern physics tells us that, at a fundamental level, the universe is made up of energy bound in fields. The physical world of our experience comes in particles and forces: the quarks and leptons that constitute matter and the four fundamental forces of nature. But ultimately even particles and forces are manifestations of underlying fields. As John Lucas put it, the word "electron" is now used as an adjective, not as a noun. Therefore, we're not talking of a something that is here or there but of a negatively charged hereabouts.

Although foundational to physics, the idea of a field is anything but simple. A field, says John Archibald Wheeler, is a physical reality that contains energy and occupies space. Einstein made the point that there is no radical distinction in quality between matter and field. A matter field has a far higher concentration of energy than a matter-less field but this is a quantitative, not a qualitative, difference. Moreover, says Wheeler, a vacuum is free of matter, but not free of fields since a vacuum has energy.

Classical physics, the study of aggregates of atoms, and quantum theory, the study of subatomic phenomena, each have their own version of field theory. Classical field theory tells you about the state of affairs at each point in space-time, e.g., the temperature. Quantum field theory tells us that space is filled with quantum fields and concerns itself with the excitations of the fields that manifest three of the fundamental interactions or forces: strong, weak and electromagnetic.

Michael Faraday and James Clerk Maxwell first introduced the concept of field in science. A force can be exerted between two physically separated bodies, e.g., iron filings around a magnet, because there is some physical agency acting in the space between them. These agencies or lines of force include gravitational, electric and magnetic fields. Maxwell's famous equations describe the behavior of electric and magnetic fields. Every location in a given field has a value such that a particle present there will experience a certain force. Disturbances of a field spread out as waves.

When quantum physics was applied to the field theory of electromagnetism, the resulting quantum field theory gave a new understanding of particles. All particles are manifestations of quantum fields, and all talk of particles should be understood as talk about the interactions and properties of quantized fields. The quantum particles in an electromagnetic field are photons. Electrons and other elementary particles have their own quantum fields, e.g., the electron field. The excitation of a field produces different quantum particles. Ripples of energy and momentum in a field are converted into particles. Fields are described by their characteristics, such as the mass, momentum and spin of an electron field, and these characteristics have definite values called quantum numbers.

Quantum field theories describe three of the four forces, strong, electromagnetic and weak, and these form the Standard Model. Thus far, it has not been possible to describe gravity with a quantum field theory. The Theory of General Relativity describes gravity but does not address the quantum domain.

The vacuum is a very important idea in the quantum world. Far from being absolute nothingness, it has its own structure and teems with force fields even at a temperature of absolute zero. It is active, not empty or inert as traditionally thought, because its quantum fields can never be still. The energy density of the vacuum is called zero-point energy. The Heisenberg Uncertainty Principle as it relates to energy and time specifies that no quantum system can assume a permanent energy state, which means that a quantum system cannot have zero matter/energy. The Heisenberg Uncertainty Principle requires random fluctuations of the zero-point energy in a vacuum. Through such fluctuations, the energy is converted into pairs of particles, say electrons or photons, and their antiparticles. These are called virtual particles because they have only a momentary existence before canceling themselves out. In *Nothingness: The Science of Empty Space*, Henning Genz notes that there is no absolutely empty space, since even in the emptiest space permitted by the laws of nature there are energy levels about which the energies of fields and particles fluctuate.

ॐ

Light

Of all the forms of energy in our world, light is arguably the most fascinating. Just think about this: at every instant of our lives we're surrounded by photons speeding by at 186,283 miles per second. These are the photons that chemically react with our retinal cells constantly to give us sight, and with chlorophyll to generate the energy required by plants, and us. And it is light that gives us the glory of color. Intelligence, energy, beauty—all at once!

Visible light is one among several forms of electromagnetic radiation that range from gamma rays to radio waves. It is made up of photons that act as waves or particles depending on the circumstance; thus, light is seen as a wave in empty space, a particle when it encounters matter. This is the wave-particle duality of light made famous by Einstein. Photons have no mass or charge and travel faster than anything else in the universe; in one year, a photon travels some six trillion miles. It cannot be split into smaller particles. Photons dating back to the origin of the universe are still with us today.

The power and intelligence of light manifests itself most clearly in modern technologies ranging from the laser to fiber optics.

The laser, first invented by the Nobel Prize winning physicist Charles Townes, is a process whereby the intensity of light or some other form of electromagnetic radiation is amplified as it passes through atoms of a specific solid, liquid or gaseous medium. The interaction of photons with excited atoms creates an avalanche effect. The resultant beam, which is coherent because various wavelengths of light are forced into one wavelength, is powerful and highly focused. Lasers have found numerous applications from printers and CD players to high-precision surgery and checkout counters. The laser is not simply a human invention: its underlying principles are operative in nature. For instance, an infrared laser was found near the star MWC 349 and an ultraviolet laser near the nova-like star Eta Carinae that is millions of times brighter than the sun.

Light is also today's most potent medium of information transmission. Information is transmitted via photons rather than electrons, so that light signals with various pulses, like the 0s and 1s in computers, are sent across an optical network. Currently a single optical fiber, also known as a light pipe, carries tens of billions of bits a second. If performance levels increase at today's rates, one fiber could carry hundreds of trillions of bits a second in another decade. Various wavelengths of light, each carrying gigabits of data, are combined and put on a single fiber. Optical amplifiers boost these signals to keep them going, and optical switches route various wavelengths to the end-user.

It may be mentioned here that light of the right color is essential for the sustenance of life. Grass is green because of chlorophyll, the molecule that absorbs sunlight to perform photosynthesis. But chlorophyll can only absorb sunlight if that light has the precise color it happens to have. Thus this process, so fundamental to plant and therefore animal life, takes place only because sunlight has the precise color it does have.

Talking of color, if we can attribute genius to a painter who mixes and matches colors to give us a great work of art, why should we hesitate to recognize a manifestation of intelligence in the color coordination of the universe? Our immediate and instinctive response to the vast and intricate variety of factors and forces that work together to produce even the smallest display of color is awe in the face of the underlying color code. The natural sciences tell us that color emerges from light and its action both on material substances and our sensory structures. In other words, the underlying scaffolding of physical reality is responsible for color. But this structure, beginning with light, the energy that comes to us from the very origin of the universe, with its various wavelengths, is just as breathtaking as its effects. Knowing that the sky is blue because atmospheric molecules scatter the shorter wavelengths of light or that a rainbow results from water droplets refracting light no more answers the question of the ultimate origin of color than a radio by itself explains the origin of the songs we hear on it.

The properties of light, such as its incredible speed that is the speed limit of the universe, its indispensable function as an energy source, its essential role in revealing the world to us, its limitless capacity for information transmission, and its production of color, are mysteries that demand explication. We are blinded not by darkness but by an excess of light when we say, "In His light we see light."

☙

Perpetual Motion

One of the great discoveries of modern science is the revelation that everything is in motion. For instance, anything on earth is hurtling through space at over a thousand miles an hour. If you stand still or sit down, like it or not, you will continue to rotate around the sun, and the sun itself will be rotating around a galaxy that in turn is racing away from the other galaxies. Moreover, your body's particles are whirling around ceaselessly, and they continue doing so even when you're dead!

So how did this motion originate? We know that the earth moves because of gravitational force: the sun pulls the earth, and consequently, the earth orbits it. Gravity itself is the result of the curvature of space, and space curves because of mass. But why does mass cause the curvature of space? Currently there is no answer to this, and even if a scientific explanation were to emerge, we could then ask why that particular state of affairs exists. The same is true of motion caused by other physical forces, such as the expansion of the galaxies from the Big Bang and the motion of an electron around the nucleus. In all these cases, we come to a point where we have no further explanation for the way things are, other than to say that's the way they are.

So There You Have It

The mystery and wonder of the world are revealed to us at three levels: (1) the very existence of a universe with components that mysteriously obey certain laws; (2) the intelligence embedded and active in the universe; and (3) the energy that mysteriously keeps all things active and manifests itself in myriad forms. The only satisfactory explanation for the energy in the world is a Source of infinite-eternal Energy. The universe is an invention of infinite intelligence and **power**.

Chapter

The Wonder That Is Life

*I*MMEASURABLY MORE MYSTERIOUS and magnificent than particles and forces and the universe itself is the phenomenon of life, of autonomous agents that intelligently process information. Paradoxically, the very pervasiveness of life has dulled us to its grandeur and glory. But even the most preliminary review of the nature and essence of life will open our eyes to its irreducible splendor.

Contemporary studies of the origin of life say next to nothing about the nature of life. There is a complete and consistent failure to consider the phenomenon of life as a phenomenon. But only a keen understanding of the nature of life can help us make progress in determining the ultimate origin of life.

So what is life? As with any fundamental datum, life can only be described in terms of its characteristics, and these are of two kinds, biological and ontological. If the *biological* is the pixel-level study of the images on your computer screen, the *ontological* is a study of the messages and stories communicated by the images. Just as pixels can explain the physical images they constitute but not the meaning conveyed by these images, so also the biochemistry of life is radically distinct from its ontological characteristics.

On the *biological* level, life is characterized by such activities as nutrition, metabolism, growth, response to stimulus, self-motion and replication, and the complex interplay of nucleic acids and proteins.

On the *ontological* level, life adds a wholly new dimension to the material universe with its hierarchy of autonomous agents driven by intelligent symbol processing through the medium of DNA (deoxyribonucleic acid) and RNA (ribonucleic acid).

When speaking of the origin of life, most people are referring to the physico-chemical precursors of biological life. But physics and chemistry cannot *describe or comprehend* the idea of existing as an autonomous agent and of processing intelligent messages using chemical codes. And if they can't *describe* this phenomenon, they certainly can't *explain* its origin. When we are talking of the origin of life, we are not talking of how certain chemicals organized themselves to form a living system. We are asking how autonomous, intelligent agents could come to be in a universe of undifferentiated matter.

In studying life as a phenomenon, we are struck especially by three of its features: autonomous agency, intelligent processing of messages, and a hierarchy of distinct forms. Once we've considered these three dimensions, we will have some idea of the nature of life and, consequently, a sounder basis for studying its ultimate origin. It seems obvious to me that by their very nature, the dimensions of autonomous agency and intelligent message processing could not emerge from matter given any amount of time or any external condition. (Can we seriously believe that the Rockies or the Himalayas could one day come to life or start thinking?). Thus, the idea that life can be reduced to matter is simply incoherent. Like the origin of material being, the origin of life is a mystery that cannot be resolved at the level of matter.

We will now consider the three dimensions of the miracle of life: (1) intelligent message processing; (2) autonomous agents that independently replenish and replicate themselves, a dramatically new phenomenon in the history of the universe; and (3) diversity that extends from cyanobacteria to dinosaurs.

ᐯᐱ

Intelligent Message Processing

At a fundamental biological level, life revolves around cells, proteins and the nucleic acids DNA and RNA.

Each cell contains 10^{12} bits of information, and any one cell has the information coded within it to build a copy of the organism's whole body. Every cell is a high-tech factory with complex transportation and distribution systems and huge libraries of information. E.J. Ambrose has classified the intricate structures within a cell under five heads:

+ The powerhouse whereby the cell receives and regulates a continuous flow of energy, for instance, energy in the form of photons from the sun passes on to the electrons in the chlorophyll molecule. The energy is distributed around the cell by energy-carrying molecules called adenosine triphosphate (ATP).

+ The machine tools—all the activity in the factory is run by tools called proteins (more on this below).

+ The raw materials—the factory uses elements like nitrogen and sodium that are passed through various production lines till the required end product has been generated.

+ The factory gates—the factory is protected by a gate of sugar molecules and a gate keeper made up of fat molecules for letting in raw materials.

+ The fertilizer plant—the factory requires ammonia or nitrates to function. The earliest microorganisms were able to attach hydrogen to the nitrogen in the atmosphere.

Proteins are the basic building blocks of the cell. They are also the tools used by the cell to create the structures of life and drive all the activities on which it depends. They form nails and tendons. They act as catalysts for chemical reactions. All proteins in living beings are made from different sequences of just twenty organic molecules called amino acids. Proteins have the extraordinary ability to assemble themselves without external intervention. This self-assembly is a process called protein folding whereby a given sequence of amino acids forms a specific three-dimensional structure. It becomes a particular protein with a precise structural or functional role. Gerald Schroeder points out that every cell in the body, other than sex and blood cells, makes two thousand proteins every second from hundreds of amino acids. This process is so complex that, as reported in *Scientific American,* a supercomputer, programmed with the rules for protein folding, would take 10^{127} years to generate the final folded form of a protein with just 100 amino acids. *But what takes a supercomputer billions of years takes seconds for real proteins.*

The living blueprint that constructs and maintains cells and determines the sequence of amino acids in every protein is DNA . DNA is the biological language of life, and its letters are Guanine (G), Adenine (A), Cytosine (C), and Thymine (T). It is a data repository of genetic information that transmits hereditary characteristics to future generations. Copies of the information encoded in DNA are transmitted via another nucleic acid called RNA to the assembly line where proteins are being manufactured. The information systematically supplied by DNA is used to construct, synthesize, repair and replicate. The DNA present in each cell contains three and a half billion nucleotide bases and, says Schroeder, has more information than three complete sets of the *Encyclopedia Britannica.*

The interaction of DNA, RNA and proteins can only be described as intelligent processing via complex symbols. As David Berlinksi points out in *The Advent of the Algorithm,* each molecule of DNA contains information in a set of signs that have their own meaning. This algorithm in biochemical code is the program that runs the operations of the cell and drives its development. While the chemicals carry the code, it is the message embodied by the code that is ultimately important. In *The Hidden Face of God: How Science Reveals the Ultimate Truth,* Gerald Schroeder shows that chemical laws may explain the bases, sugars and phosphates of DNA, but not its information content or the intelligence driving it.

Autonomous Agency

What is it that does the replication or adaptation? What responds to stimulation or ingests and excretes? What is it that exercises the intelligence intrinsic to a living system? What is the center of all the action, the moving force? Since each and every living being acts or is capable of action, it can rightly be called an agent. And since these agents are capable of surviving independently, they are therefore autonomous agents. Moreover, every such agent is intelligent because all its activities depend on the intelligence of DNA.

There are three realities at work in every living system: the intelligent agent that operates the system, the computer language used for the execution (DNA), and the hardware infrastructure (the specific combination of cells).

Every intelligent agent, i.e., every form of life, is fundamentally purpose-driven. That is to say, all of its activities are driven to meet specific goals, whether it is to nourish itself or reproduce, and the agent does this for itself, on its own. It has its own inbuilt dynamism; it is the source and cause of all its actions, e.g. growth, maintenance, and reproduction. And life, as such, does not exist independent of the intelligent agents that possess it.

The Hierarchy of Life

There are many kinds of intelligent agents and it is evident that they form a hierarchy:

1. Unicellular organisms that include cyanobacteria and other microbes,

2. Plants—these are intelligent agents because they perform all the actions characteristic of living beings, including the use of molecular signaling systems,

3. Animals—here there is a further hierarchy centered around varying capabilities for locomotion, reproduction, instinct, the operation of the senses,

4. The human person who is self-conscious, rational, free.

But from the lowest to the highest, every kind of intelligent agent has the same basic but sophisticated set of instructions in code—DNA. This instruction-set manifests itself differently on the various hardware platforms that make up the hierarchy of life.

☙❧

Intelligence in Nature

The heart of the materialist and neo-Darwinist rejection of intelligence in living beings, and Nature in general, is simply this: if you break everything down to the subatomic level, all we have left are interactions of mindless fields. At the most basic observational level there's nothing else taking place but disturbances of quantum fields. Sure, there's purposive-type behavior, and sights and sounds that evoke pleasant sensations. But when you come right down to it, you cannot show that there's anything other than fields, and the conclusion is that matter is all there is to the matter. No intelligence, purpose, beauty—just energy fields.

So here's my response. The mistake of the materialist is to assume that intelligence is something observable in the same sense and at the same level as molecules and fields, and to conclude that it's non-existent if non-observable. Everyone agrees that the activities of fields cause changes that trickle up to the macroscopic level. But how is it that there is such a relationship between subatomic and atomic realms? How is it that there are different levels in the structure of material beings? How is it that there are different properties at each succeeding level? How is it that new properties start appearing in the world?

Here the neo-Darwinists might chip in to note that, whatever the mechanism, all existing and new properties are acquired purely through material media. They point out that the sequence of adaptation and mutation that leads to all living forms today is well documented.

But we ask, then, how is it that quantum fields enable the progressive addition of properties? Is this an inherent capability of the fields or something that's added on from outside? And shouldn't the nature of the properties themselves be present in the field?

Since we're talking abstractions, a concrete instance might help clarify the issues.

Let's take a bird, an appropriate example, given that Darwin developed his theory of evolution from his observation of finch populations. At one level we observe the singing, flying in formation, navigational skills and nesting habits. At the next level down you have the feathers and beak and legs. Below that are the various internal organs and processes, bones and muscles, the digestive system, lungs, vocal cords. Further below that are the cells, and below that the molecular and atomic structures and, finally, the sub-atomic space made up of fields.

Let's see what we can't deny. We agree there's a bird. We agree that it can sing and fly. We agree it can do these things because of its vocal cords and wings and physiological systems.

My question is this: how did the quantum fields at the most fundamental level yield what we have at the top? The bird with its singing and flying is no illusion. You might say it's the fields in their particle manifestations doing the singing and the flying and you're right at a certain level. But what gave the fields this power and what turned them on to end up as beautiful birdsong and a bird in flight? When the bird dies, presumably the fields are reorganized. But certain properties are no longer with us: the warbling and twittering, the flying and nesting, and...the living. We realize that this is a different state. But how are such transitions possible? And how did it all start?

Now the properties of a bird are real. But how is it that matter acquired such properties? There are two options: matter was outfitted with these properties or the properties were progressively added.

The neo-Darwinist can give a systematic account of how birds came to be the way they are. The most popular view, displacing the crocodile-as-ancestor notion, is that dinosaurs evolved into present-day birds. Fossils of dinosaurs that appear to have feathers, such as *Microraptor gui*, are cited in support. The prototype in this paradigm is the 100+-million-year-old archaeopteryx that, although it wasn't capable of actual flight, had wings and feathers, teeth, forelimbs and a tail. Critics like Storrs Olson, curator of birds at the Smithsonian, charge, of course, that the fossil evidence for the theropod origin of birds is sketchy. At any rate, fossils that actually look like birds date from the early Cretaceous period. By the Tertiary period, about 60 million years ago, various modern birds from waterbirds to songbirds turn up. In all, some 10,000 species of birds are found on earth today. It's believed that the diversity in birds was largely caused by ecological factors. Various theories have been advanced to explain flight in birds. According to some scientists, flying started when tree-dwelling animals that could glide started flapping their wings, while others think it started with small, predatory dinosaurs that used to run with arms outstretched and leap into the air.

While this kind of a history may be useful as far as it goes, it leaves the primary question entirely unanswered. And that's how matter allowed all this to happen. Ecological factors may have unlocked the capabilities of the organisms in question, but where did these innate capabilities, the unique flight feathers with its aerodynamic capabilities, the respiratory system found in no other mammal, come from? And how is it that such adaptation to ecology was possible? Take it down to the genetic level: even with random mutations, the genetic material must ultimately have contained the capabilities we now see manifested in sparrows and robins, meadowlarks and woodpeckers. Or—if you go back before the origin of life—you ask how the matter fields existing before the origin of life acquired all the capabilities that were to manifest themselves after the genesis of life and living beings. Were those innate capabilities, and if so, how did matter come to have such properties, and how is it then able to manifest them?

The point I'm trying to make is that there's something involved here beyond quantum fields, right from the very beginning. When we see birds take off with a runway of inches or fly across continents for months at a time, we have a right and an obligation to be wonder-struck. To lightly dismiss this as a product of evolutionary adaptation is to ignore the questions of (a) where did the entity that does the evolving come from, (b) how did it come to have its innate capabilities, and (c) why is reality structured so that adaptation is possible.

In view of the widespread confusion surrounding these issues, I feel compelled to offer an analogy. An artist mixes certain colors and starts painting away in her studio. After several days of thought and labor, we are shown an exquisite work of landscape art.

What is it that we see? Now if we look at the painting, at one level it's entirely a matter of paints thrown together on a canvas surface. At a more basic level you have colored particles in a liquid medium. At a chemical level, you have organic and inorganic pigments, solvents and binders made up of organic molecules called monomers that combine to form polymers. At a subterranean level, you're back to fields again, so that the painting at all stages is simply a matter of fields going through different interactions. So should we say that fields are the whole story from beginning to end? Or is there a painting, a landscape, there that's a new reality? Granted, the painting at a fundamental level is just a matter of quantum fields, and the process of painting too is simply a series of interactions undergone by the fields. But we're obviously missing something here. *What we're missing is the whole story.*

It's not that we don't see the forest for the trees. We don't see the trees either.

But what's there in addition to the pigments, binders and solvents? First, there's a picture for which these chemical components are a vehicle. Second, the picture was created from the intentions and actions of the artist. A purely physico-chemical discussion of the painting will not and cannot say a word about either of these two dimensions. But no one has any doubt that both these dimensions are just as important as the physico-chemical side of the story. Moreover, all of the physical changes and transformations are driven by the intentions and actions of the artist, and they *only make sense in the light of the end product.*

From simply analyzing the various constituents of the paints, we could never have guessed that they'd end up giving us the ultimate configuration we're left with, a work of landscape art. If you analyze the components of the painting, all you have are certain chemicals lumped together. Even the history of the painting only shows these chemicals progressively moving through random configurations. At the most basic level we're left with the activities of fields.

The materialist might admit there's something new. But it's all simply a matter of material interactions of pigments, binders and solvents, albeit driven by material interventions of the painter. There's nothing over and above matter.

In one sense it's only matter. But obviously this is not all there is to it. We're not merely being metaphorical when we say that there are other equally relevant and valid levels relating to the painting that must be grasped.

This we know:

- ◆ The mind of the painter played the main part,
- ◆ Her efforts and interactions were essential,
- ◆ A new thing was brought into being from the interaction of mind and existing material.

The painter had something in mind, and what she had in mind she made a reality in the resulting picture. The chemical components were a vehicle for the picture, and by extension, for the artist's intentions and actions. The mind of the artist had the decisive role, and it was the operation of the mind that gave us the picture and the progressive addition of new properties. But if you stayed simply at the level of pigments and particles, you'd never know anything about the painting or the intelligence at work.

This is precisely what's involved with the intelligence in nature. At a physico-chemical level, nature is strictly fields and their particles, atoms and molecules, genetics and adaptation. But at the macro level these basic components become vehicles of rationality and beauty and purposiveness. Only a confusion of categories would lead us to look at these higher-level attributes with lower-level tools. Schroeder highlights the paradox: "In one mix of protons, neutrons and electrons I get a grain of sand. I take the same protons, neutrons and electrons, put them together in a different mix and get a brain that can record facts, produce emotions, and from which emerges a mind that integrates those facts and emotions and experiences that integration. It's the same protons, neutrons and electrons. They had no face lift yet one seems passive while the other is dynamically alive."

Intelligence is a hard fact of living things and Nature as a whole; that is, if you think the laws of nature, the purposiveness of replication and the grandeur of energy, to name just three examples, count for anything. When you see a painting, by all means admire the polymers and pigments, but don't remain at the ground level. It's like saying that *Macbeth* is really just black marks on white paper or a collage of costumes, sets and people saying certain words from memory. This is not to see what the play really is. It is to entirely miss the play.

So the next time you see a bird, don't view it simply as a twig on the evolutionary tree, a vehicle of natural selection or matter in motion. Consider its displays of intelligent, even extraordinary, behavior, and assess these displays on their own terms. Listen to it sing. This is not just a disturbance of fields, although it is that, or phonons gone berserk. It's a bird singing. And it's here, now. Its existence is as marvelous and moving as the most beautiful painting. Yes, it emerged through progressive modifications from primordial times, but the endgame is what matters, as surely as the end painting makes sense of the initial pigments. Its properties of intelligence and beauty bespeak a source, in the ultimate analysis, that is intelligent and beautiful. Like the painting, it most certainly is a manifestation of matter fields, and like the painting, the form taken by the manifestation cannot be explained without reference to mind.

The Origin of Life

We have said that the question of the origin of life can only be answered in the context of its nature. This nature is not simply a matter of proteins and acids, important as these may be, but of the extraordinary phenomenon of autonomous, intelligent, self-replicating agents.

Current scientific investigations into the origin of life focus on the question of when and in what setting life first appeared. It is certainly of cardinal importance that we try to discover the primordial conditions and physical templates in play at the starting point. In tandem, we will want to determine the chemical processes present before and after the genesis of this sophisticated energy-processing self-replicating system. Clearly life had to have a beginning and this beginning would have a biochemical dimension to it. But the origin of life was not just a biochemical happening; the biochemistry was driven by what gives every appearance of intelligent agency.

The earliest fossils of life, fossils going back some 3.85 billion years (recent studies place definitive evidence at only 1.9 billion years ago), are of modern photosynthetic cyanobacteria with all the capabilities of life from metabolism to reproduction. How life with all the complexity of DNA came to be so suddenly after the formation of the earth is a great mystery.

At the biological level, we still have no conclusive answers. John Maddox, the former editor of Nature, notes in his book *What Remains to Be Discovered: Mapping the Secrets of the Universe, the Origins of Life, and the Future of the Human Race* that two major problems in biology have been virtually untouched, and the origin of life is one of the two. The problem is exacerbated at the ontological level where progress is possible only if we recognize life for what it is. The cyanobacteria that are the first known living beings were intelligent agents that were the source and cause of their actions with goal-driven, inbuilt dynamisms of self-generation and information processing. No geological or biochemical study is going to explain even the nature of such autonomous intelligent agency, let alone tell us how it could come to be.

When we think of the origin of life, we are really talking of the origin of living beings since there's no abstract thing called life. A living being, be it a fern or a dinosaur, is an autonomous agent that processes information, generates energy and reproduces using the incredibly intelligent symbol-processing system we call DNA. Such a being is not simply a configuration of matter or a life principle. It is a new kind of reality, the reality of being an agent and functioning as an intelligent system. When you take the chain of life back to the beginning, you ask how the first living being came to be. To answer this you have to know what it is. Since life is a reality of autonomous agency and intelligent processing, we realize that such a reality can only come to be if it comes from a source that is not just intelligent but also an agent.

Thus we are led to the conclusion that life of any kind cannot come from non-life. In the ultimate analysis life must spring from a transcendent Source that is not simply living but is Itself Life in all its fullness. Dr. Werner Arber, winner of the Nobel Prize for the discovery of restriction enzymes, lays out a compelling case for this train of thought:

1. Life only starts at the level of a functional cell;

2. The most primitive cells require several hundred different specific biological macro-molecules;

3. It is a mystery how such already complex structures came together.

4. The possibility of a Creator, of God, represents a satisfactory solution to this problem.

Chapter

Homo "Sapiens"

*T*HE GREATEST SUPERSTITION of the last 200 years is the widespread idea that the conscious thinking experienced by all human beings at every waking moment was produced entirely from and by mindless matter and is in fact reducible purely and simply to matter. We have lost our minds in more senses than one!

Of this we are all certain: we are conscious and aware that we are conscious. The nature of data processed by our consciousness is wide-ranging:

- ◆ **sensations** that reach us through one or more of the five senses
- ◆ **memories** that "relive" sensory experiences or sensations of the past
- ◆ **images** that we form in our imagination by extrapolating from our sensory experience
- ◆ **concepts** that do not have any correlation with our sensory experience such as the notion of liberty or mathematical entities and theories
- ◆ **intentions** that we form and execute such as planning to go for a walk or a vacation

+ **choices** that we make ranging from giving up our lives for our country to telling the truth in a conversation.

Some of the data of our consciousness are directly related to the physical world, for instance, the things perceived by the five senses or entities that we imagine from our previous sensory experience. Others are simply pure acts of the intellect, for instance, the acts of understanding, judging and reasoning.

Consciousness

We are conscious and aware of being conscious and, in addition, we are just as clearly conscious that our consciousness is dramatically different from anything material or physical. We know that thoughts and feelings do not have physical properties such as size or shape. We know that our mental activities are accompanied by physical processes and also that we cannot see a thought if we open up the brain. We may see neural activity, but that's not the same thing as seeing the thought *in the mode we experience it*. Conscious thought has a reality of its own that cannot be perceived as being physical in any relevant sense of the term.

In a public discussion between Richard Dawkins, the leading Darwinist critic of religion, and Steven Pinker, the foremost contemporary writer on the brain, Dawkins admits:

> There are aspects of human subjective consciousness that are deeply mysterious. Neither Steve Pinker nor I can explain human subjective consciousness—what philosophers call qualia. In *How the Mind Works* Steve elegantly sets out the problem of subjective consciousness, and asks where it comes from and what's the explanation. Then he's honest enough to say, 'Beats the heck out of me.' That is an honest thing to say, and I echo it. We don't know. We don't understand it.

Says Ian Tattersall, curator of anthropology at the American Museum of Natural History, "While we know a lot about brain structure and about which brain components are active during the performance of particular functions, we have no idea at all about how the brain converts a mass of electrical and chemical signals into what we are individually familiar with as consciousness and thought patterns."

When we perceive a neural firing in the brain, we perceive a physical process, rather than that which we experience on the inside as a thought. The argument that the thought and the neural firing are identical although they look and feel different quite clearly flies in the face of the only empirical evidence that will ever be available to us. This evidence is of two kinds: what we can know about the physiological workings of the brain and the sensation or thought of which we are directly and immediately conscious. Consciousness, as we experience it, is irreducibly trans-physical although it interacts constantly with the physical. Most important of all, it's thoughts that drive the corresponding neural transactions and not the other way around. *There is no chicken or egg question here. The thought comes first and, as a result, causes certain brain events.*

Interaction between the mind and the brain doesn't by any means imply that the brain and the mind are the same thing, just as the conductor is not identical with the orchestra. Damage to the brain obviously affects the functioning of the mind, since in its present state the mind works with the brain. When its batteries run out, the radio can no longer transmit the messages it used to pick up. The physical components of the radio, however, are by no means identical with the messages, let alone with the person or persons from whom the messages originate.

Two other fundamental features of our experience are of special importance here: our use of language and our self-identity.

Language

Syntactical language is unique to human beings. It's found even in ancient civilizations and instinctively mastered by children at a very young age. Although animals communicate through various sounds, we know that they are incapable of forming verbal or grammatical structures. Chimpanzees tutored by humans learn visual signals and then use these signals to get things. But this learning is really a process of being conditioned to perform certain actions in response to given stimuli, something that's qualitatively different from such innate human capacities as seeking explanations and deducing relationships of cause-and-effect. Nor can the chimps manipulate the symbols they learn to give new meanings as children do. The words they learn are always derived from physically experiencing the thing designated by the word. Human children, with their power of conceptual thought, do not have to learn the meanings of words from perceiving the things to which they refer. They can be told what something means just by explaining it in terms of something with which they are already familiar. For example, once they know what animals are, they can understand the notion of a place where animals are kept without having to actually visit a zoo.

Journals like *Scientific American* admit that scientists cannot explain the origin of language or the jump from primitive to syntactical language. With "the arrival on Earth of symbol-centered, behaviorally modern Homo sapiens," writes Ian Tattersall, "an entirely new order of being had materialized on the scene. And explaining just how this extraordinary new phenomenon came about is at the same time both the most intriguing question and the most baffling one in all of biology."

The greatest theorist of language in modern times, Noam Chomsky, holds that the capacity for language and especially the complexities of syntax can only be explained as innate in humans. Moreover, as David Braine points out, "linguistic understanding and thinking in the medium of words have no bodily organ through which they operate and no neural correlate." We don't know what linguistic statements we're going to make in the future, but we're still able to produce and understand them. There's no correlation between language and brain states.

The "I"

What is it that perceives and conceives, feels and thinks, judges and chooses? It is the self, the center of our consciousness, and the unitary unifier of our experiences, which gives us the identity of being the same person throughout our lives although the physical components of our bodies change constantly. Obviously, each person changes in response to different experiences, develops habits as he or she keeps performing certain acts, and so on, but there is no question it is one single person to whom these changes occur. Although we are not conscious of the self separately from its acts, we are conscious of it as the ground that pervades and unifies our acts, the entity that thinks, wills and feels. When Descartes said, "I think, therefore I am," he already assumed that his self existed. It is *I* who think, and the thinker cannot be separated from the thought.

Although we can associate specific brain events with various mental acts, these correlations do not include the most fundamental mental reality, the "I" who is conscious and does the thinking. The "I" is not present in any region of the brain. Nevertheless we cannot seriously deny the existence of the self that is "I." Remember the story of the student who asked her professor, "How do I know I exist?" He replied, "And who's asking?"

Speaking of Descartes, it is sometimes said that he regarded human persons as disembodied egos. We cannot emphasize enough the fact that our experience of the self is of an embodied self. The self is a unity of mind and body. All of our sense-data come from the body, and the mind is required to process the sense-data. Nevertheless, the mind goes beyond sensory data, beyond sensation and imagination, to think thoughts and form concepts that have no correlation with the senses. In principle, there is no reason why the mind can't exist separately from the body, since thinking and willing are not physical in nature. However, the separation of mind from body is truly a dissolution of human nature as such, and a disembodied self exists in an unnatural state.

Be that as it may, the recognition of the self, writes David Lund, is devastating for materialism. This is so because the material world in itself, and materialism tells us that this is the only world that exists, has no centers and cannot recognize a perspective of "I," of the first person. But this "I" is that which is most obvious in our experience: it is I who am conscious, I who choose and intend. The self, like its acts, cannot be reduced to the physical.

I think, therefore God exists

Could these fundamental features of the human condition of which we're directly and constantly aware, and aware of as intrinsically immaterial, have arisen from lifeless, purposeless mass-energy given not just a few billion years but an infinite period of time? It's almost as if we were to say that given an infinite amount of time, a pen and a paper, without any external intervention, would somehow give rise to the concepts embodied in the Gettysburg Address. We're not speaking of the words that constitute the speech, but the concepts of liberty and equality and justice represented by the words; such concepts are so radically different in nature from the physical objects used to represent them that it's simply nonsensical to suppose that the latter could produce the former.

We think, and our thinking is so obviously distinct from the physical realm that we cannot conceive of it as having risen from the physical. Certainly we cannot give any credence to the idea that a certain bundle of mass-energy just happened to exist without beginning or end, and then evolved without any direction or guidance over time into *THOUGHT.* We cannot seriously believe that intellect sprang out of mindless mass-energy, consciousness out of lifeless matter, and intelligence out of blind force fields. Once we truly recognize that we're conscious thinking agents, we can never again conceive of our coming to be from anything less than a conscious thinking agent. Only Mind can beget mind; only an infinite Intelligence, an intelligence that has no limitation of any kind, can create beings with any kind of intelligence. Descartes said, "I think therefore I am." It would have been more correct to say, "I think and therefore know that I cannot as a thinking being have come into existence from non-thinking matter." I think; therefore God exists.

Chapter

Seeing God Through the Wonder of the World

*I*CONTENDED in the beginning that it is possible to see God. Admittedly, a lot depends on what you mean by "seeing" and "God."

I will be considering seeing both in terms of physical perception and conceptual comprehension of meaning. How does seeing apply to God? Along with the wisest and holiest men and women in history, I hold that God is infinite intelligence, the fullness of being, and the plenitude of all perfections, an omnipotent and omniscient Spirit, the Creator of all things. As an infinite being, God, by definition, has no limitations, of space, of time, of any kind.

Once you understand what is meant by the concept of God, you understand immediately that the Infinite is not something you can see physically. In order for you to see something physically, it must have shape and size and color. It must reflect light and work its way through our sensory channels until it is registered in the brain. But if you apply these attributes to any being then it can no longer be thought of as infinite. Consequently, we cannot seriously expect God's face to appear on a hillside. And if the words "I exist, signed God" were to suddenly appear in the sky, this would not prove anything one way or another since there is no way to verify the source of the message.

Nevertheless, I hold that in a very concrete sense we can actually touch, see and feel God and that, in fact, all sensations, all sights and sounds are perceptions of God. Why? Because God invented each one of them and sustains and participates in them. God is not simply the Cause of all things, but a Cause that acts and is known through its effects.

In other words, we can recognize and become aware of God's existence just by perceiving the things around us. This is an act of seeing things *as* created, *as* requiring God's existence to explain its existence, *as* dependent and, finally, *as* manifesting the Infinite here and now. Just as we see a poem as a poem and not as print marks, and cannot see it as anything but a poem, likewise we no longer see the things around us simply as atomic facts but as realities that manifest and reflect God.

To be sure just as physical sight can be lost, either deliberately or accidentally, we can easily lose the sight of God. The face of God gazes out at us through all aspects of our daily experience. To see this face we must be capable of sight. If we mistreat or neglect our eyes or suffer from astigmatism, myopia or an infection, we can easily impair or even lose our sight.

Let's apply all this concretely. What do you see in Nature? Roaring waterfalls, mist-laden mountains, flowers of various hues and colors, grasslands dancing in the wind. Indisputably, Nature is bursting with activity at every instant, exuding incredible energy and following precisely structured laws. The merest stone is a universe of particles spinning around like perpetual motion machines. The cells of all living plants and animals are data warehouses sucking in signals and spitting out instructions. Once our minds have truly grasped the universal reality of innovation, intelligence and energy, we are only a few steps away from recognizing the manifestation of God in and through the intricate order and the ceaseless hum of activity underlying all things.

Every perception of the senses is a perception of something that was quite literally invented and sustained in existence by God. To perceive it is to perceive God because the energy, order and information present in all things—galaxies, streams, birds—are energy, order and information implanted and sustained by God at every instant. So also the unique combination of properties (shape, color, smell) that constitutes every thing was thought up and is kept in being by God. When we look at a blade of grass, we realize that its green color, its shape and smell, and the hidden engines of photosynthesis are just as much creations out of nothing as the entire universe. God is not abstract and remote but immediate and concrete. We decipher the divine language with the alphabets of this world. Once the poem reveals itself through the body of print, the print marks become both medium and message.

Now how are we to understand the notion that God continues to keep us in being? We can certainly make sense of the idea that God initially brought something into existence and set the whole process in motion. But once this is done, why is there a need, so to speak, for God's continuing involvement? Why do we need to talk of God conserving things in being?

A helpful analogy of our continuing dependence on God is sunlight. The sun is the source of sunlight. But sunlight is also constantly dependent on the sun for its continued existence. A sunray subsists only as long as the sun keeps it in being. As applied to God and us, the analogy breaks down here since sunlight is a part of the sun, whereas we're not a part of God. Also, there's an infinite gulf between God and us when it comes to the kind of existence we have. And yet our existence is far more dependent on God than sunlight on the sun. Let's see why.

Look at ourselves. At every instant we breathe just to stay alive. For us to continue living, the atmosphere must contain a certain amount of oxygen, our respiratory systems must process the oxygen, our anatomies must function as they're supposed to, and finally, the matter fields that make up our bodies must behave in accordance with certain laws. We're able to read this only because all these conditions are in place. But who keeps the underlying laws of nature in place? Ultimately it's God who ensures their continued operation. To exist is to be energized by God. But if He brought about the laws of nature, couldn't He make them so they don't need to be continually upheld in being? Which brings us to the vital part. The laws and the energy underlying all existent beings have to be constantly sustained by their author, because they're not themselves conscious or powerful or rational. They're not persons. If God were to disengage from these laws for a nanosecond, we would be instantly annihilated. This is why the famous quantum physicist Werner Heisenberg made the stunning comment, "If the magnetic force that has guided this particular compass—and what else was its source but the central order? [God]—should ever become extinguished, terrible things may happen to mankind, far more terrible even than concentration camps and atom bombs." Thus, God did not simply invent all things but constantly holds them in being.

A famous atheist I knew, Sir Alfred Ayer, once wrote that a primrose in a meadow was simply a primrose for him; he couldn't find anything transcendent about it. Similarly, many skeptics decry any attempt to go beyond the data of the senses. But this body of data is all we need, provided we bring our minds and hearts to bear on everything around us in their true and full reality. When we see a primrose we see a unique combination of colors, chemicals and communication systems that pulsates with life and rationality. God is not an additional property or add-on to the flower that we can also see if we look hard enough. Rather, *in* its uniqueness as a flower, in each one of its properties, in its life and activity, this primrose manifests God. It is not a question of whether a primrose is just a primrose but of recognizing the primrose for what it is: its uniqueness, its life, its structure, all of which express and embody the being that invented it along with all its features while keeping it together at all times.

All scientifically literate people today accept that every object in the universe is unique, constantly active and laden with information. It is precisely in becoming aware of this uniqueness, activity and information embodied in all entities that we become aware of God. Once we see all things in the light of these three features, their creativity, dynamism and symmetry, we see them perhaps as we have never seen them before. The feel of the ground under our feet assures us of its being there. The voice of our mother calling out to us is not just a matter of vibrating air molecules but embodies her in our minds. In like manner, the recognition of all beings in their uniqueness and energy and rational structure is simultaneously the recognition of them manifesting their inventor and sustainer.

Let me say here that I am not arguing for any version of pantheism or monism. There is no identification, however subtle, of God with the world, no suggestion here that God is in any sense embodied by the world or that the world is in God. We must distinguish radically between God and the world. God is not one more thing in the world, let alone all things in the world. But we do experience the transcendent in the immanent, the infinite through the finite.

The scents and colors of a beautiful rose come to us directly from God, manifesting the divine Presence and Glory. And, of course, the rose is not God or a part of God. Rather, the rose in its entirety not only reflects the glory of God, like a brilliant work of literature manifests the mind of its author, but makes God present to us as an immediate and continuing manifestation of:

- the divine power that holds it in being and sustains its activities
- the infinite intelligence that conceived it
- the ineffable beauty from which its colors and scents spring.

When we look at a beautiful work of art, we see it simultaneously as beautiful and as deriving its beauty from its creator and reflecting his or her genius. Similarly, when we see the rose we are struck simultaneously by its beauty and by its manifestation of the power and the glory of its inventor and sustainer. Moreover, unlike the artist who works with pre-existing materials like paints and canvas, the inventor of the rose starts with absolute nothingness and continues to sustain its energy of being at all times.

To see a rose is to see an immediate and concrete manifestation of infinite creativity, intelligence and energy—and that is to see God here and now.

SAGES & SCIENTISTS

*A*T THE OUTSET I spoke of three discoveries that are fundamental to our discussion. In "Seven Wonders" I considered two of them: the idea that the universe manifests the reality of infinite Intelligence and the fact that it is possible to encounter this Intelligence, that is, to see God, here and now.

The third discovery is almost as exciting. *It is the astounding revelation that the four greatest thinkers of the four major world religions, Hinduism, Judaism, Christianity, and Islam, jointly and severally, over a period of three hundred years, crafted the metaphysical matrix that underlies modern science.* This infrastructure, which we shall refer to here as the Matrix, is a meta-scientific Theory of Everything that enables scientific Theories of Everything; it is the engine of logic that underlies the scientific method as we know it; and it is the vehicle of enlightenment that keeps at bay the forces of intellectual darkness that have always threatened the scientific enterprise. It is, in a nutshell, the Big Idea:

- ◆ that the world really exists separate from us and from any deity
- ◆ that the universe is intelligible and rational because it was brought into being by sheer Intelligence
- ◆ that we ourselves are capable of knowing truths about the world be-

cause we have minds separate from matter that function rationally and discern meaning.

What is more, as we shall see, the very scientists who manufactured modern science are the *Prophets of the Matrix*. From Newton to Maxwell to
Einstein, from Planck to Heisenberg to Dirac, they testify to the truth that
the Big Idea embodied in the Matrix is the launching pad of science.

Before we go any further, a preliminary overview is in order. Every
thousand years or so, there comes a thinker whose life is as striking as
his or her intellectual output is stunning. Viewed from this perspective,
it is remarkable indeed that within a period of 300+ years, the world
was to witness the convergent odysseys of four titans of thought who
set the agenda for the study of reality at every level. This is the period I
like to call the Golden Age of human thought. Between them, Avicenna
of Persia (980–1037), Moses Maimonides of Egypt (1135–1204), Thomas
Aquinas of Italy (c. 1225–1274) and Madhvacharya of India (c. 1238–1317)
created a magnificent monument of thought that underpins the very
possibility of the scientific enterprise. It was the mother of all Theories
of Everything, one that was validated both by its inherent logic and the
success of modern science.

The point of departure for these thinkers, whom we shall call the Four,
was simply that things *exist*. From this bare fact their minds soared to the
greatest insight possible to the human mind, the realization that things exist only because there exists One who cannot not-exist, who exists without
beginning or end or any conceivable limitation. The very essence of this
Being is to BE. There is no question of was or will be for It always IS. Thus
we speak of "It" as "He who IS", the "I AM." Each one of the Four considered this equation of God to be THE fundamental truth:

> **Avicenna**: In God alone, essence, what He is, and existence, i.e. that
> He is, coincide. God's essence is to exist. "The essence of the Neces-
> sary Existent [God] can be no other than existence."
>
> **Maimonides**: "His existence is identical with his essence and his
> true reality, and his essence is his existence."
>
> **Aquinas**: "There is a being, God, whose essence is His very act of
> existing."
>
> **Madhvacharya** (Commentary on verse 17 of the *Isavaya Upanishad*

Basya): "'SO AHAM ASMI.' This is the great ineffable name of God, 'I am that I AM,' 'That Supreme Being (*asau*) which indwells in Asu is the I AM.'"

Interestingly enough, this striking idea of God as the Self-Existent is to be found in the Hebrews and the teachings of Zoroaster. "Then Moses said to God... 'But if they ask me what his name is, what am I to tell them?' And God said to Moses, 'I Am who I Am...This is my name for all time; by this name I shall be invoked for all generations to come.'" (**Exodus** 3: 13-15). "Thus spake Zarathustra—'Tell them, O Pure Ahuramazda, the name which is the greatest, best, fairest and which is the most efficacious for prayer.' Thus answered Ahuramazda...'Ahmi yad Ahmi Mazdo: I am that I AM.'" (**Avesta**, xvii, 4–6).

The great *discovery* of divine self-existence, the "God equation" of Essence=Existence that has inspired hundreds of writings, is foundational for the Matrix. From it flows a dynamic vision of reality rooted in a living, ever-active and infinitely creative source and conserver of everything that was, is and will be. By working out all the implications of this equation, the Four arrived at all their other findings:

♦ The world is real and rational

♦ The human person can think and know

♦ Every phenomenon has an explanation given that infinite Intelligence is the ground of all things.

Why is the Matrix important for science? Well, for modern science to work, for the very possibility of a scientific method that bears fruit in theory and experiment, we must make certain basic assumptions about the nature of the world. For instance, we couldn't do science in the sense of seeking out underlying causes and laws if we didn't believe that the world operates with causes and laws. Nor could we pursue our inquiries if we didn't think our minds are capable of making deductions and reaching valid conclusions.

But why should we believe any of these assumptions to be true? And how did we come up with them in the first place? Did scientists discover them the way they discovered Pluto or invent them the way they invented jet engines? The fact of the matter is that science and the scientific method didn't drop out of nowhere. There's a framework of thought behind science that goes beyond the methods of science. It's a set of pre-scientific and pre-philosophical insights accepted by the first scientists.

We call them meta-scientific and by that we mean a principle or reality that is *fundamental to science but cannot be tested with the methods of science.* The domain of the meta-scientific includes:

♦ Things that have no physical characteristics, e.g., consciousness, abstract thought,

♦ Claims that can be proved or disproved by reasoning but not by experiment, e.g., are our minds capable of knowing?

♦ Questions about the nature of existence, e.g., what does it mean for something to "be?"

A classic meta-scientific issue is the belief that the universe exists. This can only be assumed by science and not proven because every physical experiment will necessarily assume the world exists. A proof for the reality of the world, as laid out by Madhvacharya, for instance, is necessarily meta-scientific.

The Matrix does two things: It affirms the meta-scientific principles that were later adopted by science and then builds a case for accepting the truth of these principles.

To put it another way, it supplies science with its foundations and provides the ground on which these foundations can be laid. Most scientists are too busy, as they should be, building on the foundations to worry about the foundations themselves. But if we assume, as science does and must, that there's rationality in the world embodied in the laws of nature, then we should know if and why this assumption is true and what it implies. It's here that the Matrix takes us beyond the assumption itself to the ultimate reality on which it's founded.

The importance of the Matrix becomes apparent when we consider the idea of scientific laws. The notion of fundamental laws of nature is now a commonplace in science. But where did the idea of such laws come from? Not from atheists or materialists. Intellectually, it originated in the idea of a divine Mind who instituted immutable laws of nature, as even critics of the concept of laws of nature admit. Paradoxically, the scientist who today reflects on these laws talks of the Mind of God. So here are the two sequences: *historically*, the idea of God led to the idea of fundamental laws; *currently*, the idea of fundamental laws leads to the idea of God.

The phrase "Mind of God" used by scientists today is worth scrutiny. It's not simply a metaphor for the laws of nature as some have said.

First, the notion of mind is critical. A mind is the center of consciousness, intellect and will; it's the faculty that thinks, intends, plans and acts. A mind is intrinsically and qualitatively distinct from matter. If minds are material then it's not possible to have ideas or intentions, since it's incoherent to suppose that particles can have a center of cognition. Minds can interact with matter, but they do not occupy space. The Mind of God is the Mind that invented all things, and brought them into being from sheer nothingness. It is the Mind that imprinted intelligence in matter and generated the hierarchy of intelligent living systems.

Secondly, the origin of laws of any kind can't be sought in a material matrix. Matter can't produce concepts or patterns or mathematical constants. A force field doesn't plan or think. If we look for a source of the laws of nature, then we necessarily have to start with a mind that can't be reduced to matter. It has to be a mind of great acuity since it has to conceive the logic that runs a universe of hundreds of billions of galaxies. It has to be a very powerful mind because it needs to not just generate the applicable mathematical equations, but then to call a cosmos into being that follows the formulae. And, as we will see, it also has to be an infinite Mind because the existence of anything at all can be explained only if there is a being that exists without any limitation.

It is my contention, then, that the very concept of fundamental laws of nature sprang out of a vision of the world as a creation of an infinitely intelligent Mind. "There is a certain Eternal Law, to wit, Reason, existing in the mind of God and governing the entire universe," wrote Aquinas. Since laws of nature are a primary part of the scientific enterprise, it is clear that these foundations of science are embedded in a divinely grounded vision of the world. The systematic formulation of that vision is what we call the Matrix.

ෆ

The Matrix of Modern Science (Diagram)

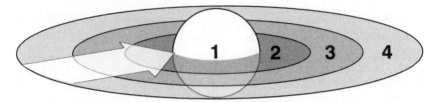

Matrix = Womb (mold within and from which a thing is formed)

1. **(Innermost circle)** Eternally-existent, all-powerful, infinitely perfect Intelligence (Avicenna, Maimonides, Aquinas, Madhvacharya)

2. **(Next circle)** The World, created from absolute nothingness by the infinite Intelligence/God (Maimonides, Aquinas), imbued by God with fundamental laws (Maimonides, Aquinas) and separate from God and ourselves (Madhvacharya)

3. **(Third circle)** *Homo sapiens*, a union of matter and spirit (Avicenna, Maimonides, Aquinas, Madhvacharya) with a mind capable of transcending matter by reasoning, understanding and finding truth (Aquinas, Madhvacharya)

4. **(Fourth circle,** with arrows pointing to circles 2 and 3) Modern science, built on the beliefs that (a) the world exists separate from our minds and a deity; (b) the world follows fundamental laws and is therefore rationally ordered; and (c) we are capable of reasoning, understanding and finding the truth about the world.

At this point, let me anticipate the most obvious questions and critiques:

♦   What evidence is there for such a Matrix and, equally important, why should it be attributed to just these four thinkers?

♦   Why should the works of precisely this particular group of thinkers be clubbed together? What rationale lies behind the selection?

♦   Is the Matrix written up as a document? Is it embodied in a given text? Or is it just a fuzzy way to classify disparate ideas?

♦   What evidence is there that the purported Matrix influenced the birth of modern science in any way, and specifically, what evidence is there that these thinkers had anything to do with the progress of science or that scientists consciously thought of the Matrix?

♦ The three components of the Big Idea seem to be glimpses of the obvious and there is no need to invoke a mystical Matrix as its source or fountainhead.

♦ The growth of science was spurred by progressive liberation from the shackles of religion. Science was the source of enlightenment and religion of intellectual darkness. Isn't it a bizarre move then to tout four religious thinkers as the standard bearers of science?

♦ Hasn't philosophy always been an obstacle to science?

♦ Hinduism, as generally understood, says that the world and we are a part of God. And the notion that the world or we are real is simply an illusion. So why introduce Hinduism as a part of the pantheon?

♦ A number of modern writers, Fritjof Capra comes to mind, have popularized the idea that Eastern mysticism and philosophy is the true counterpart of modern physics. Where does this fit in with the Matrix idea?

♦ Who determines who are the four "greatest" thinkers?

♦ If certain parts of the so-called Big Idea are true ("the world is real"), there is no logical reason to conclude that all elements of it are valid ("God exists").

All of these are legitimate concerns. But the discovery of the Matrix is novel and exciting precisely because it gives unexpected but unexceptionable answers to these very questions. The succeeding sections of "Sages and Scientists" build my case as it relates to most of these issues. I will, however, start off by responding in summary fashion to these questions.

Evidence for the Matrix

The evidence for the Matrix is found in the writings of the Four themselves. With the possible exception of Madhvacharya, their works are found in most libraries. David Burrell's *Knowing the Unknowable God: Ibn-Sina, Maimonides, Aquinas* lays out the common themes and organic progression in the thought of three of the four thinkers. I will be outlining key ideas of all four as it relates to our discussion. Certainly the set of ideas laid out in the Matrix predates the Four. It has been fleshed out in the writings of many other thinkers and in the scriptures of Hinduism, Judaism, Christianity, Islam and other religions; it is, arguably, also the fountainhead of the most primitive religions. I have singled out the Four because they articulated, systematized and refined trains of thought that had previously been disjointed and indistinct. They created a viable and potent framework that crystallized the collective wisdom of humanity. Although all four had their distinctive approaches and emphases, their exposition of the central themes and their defense of this position against its antagonists established an intellectual beachhead that, in my view, remains in place to this day.

Why This Particular Four?

Each one of them brought something singular and indispensable to the table. Avicenna first articulated the distinction between essence and existence and the unity of both in God. Maimonides refined this idea further and introduced the key notion that the properties of the universe derive from God and are therefore immutable, although God can change them. Aquinas moved the essence/existence discussion to its zenith with the insight that God's essence is his very act of existing; in him all potentialities are actualized. Aquinas also noted that the laws of the universe proceed from the Mind of God. Madhvacharya showed cogently that God is distinct from the world, and provided the most powerful refutation of monism in history. Aquinas and Madhvacharya also gave unique and enduring responses to the all-important question of how we can know what is true.

The Physical Embodiment of the Matrix

The Matrix isn't laid out in a single document, like, for instance, the Magna Carta. It's a description of a definite way of viewing the world as embodied in the writings of the Four. We will outline in more detail the central themes held in common by the Four and the rationale they developed to support these. So why club them all under one category? Well, look at quantum physics. Is the Quantum Idea a document or one particular text? No, it's a collection of diverse ideas generated by a number of scientists over a period of time. There are certain themes in common and various ways in which these are expressed. Three different mathematical formalisms were developed to prove the same point. But it's certainly quite acceptable to classify the whole enterprise under a phrase such as quantum theory or the quantum idea. So it is with the Matrix.

The Matrix that Gave Birth to Science

It seems indubitable that science rests on a conceptual matrix. In other words, it assumes that certain fundamental ideas are true. It is these ideas that are foundational for the scientific method. Science assumes that:

+ The material world is real, not an illusion

+ Different things are distinct from each other

+ There are interactions between these things that can be measured

+ There is an explanation for every phenomenon, or a cause for every effect

+ There are regularities in nature that can be captured in laws and mathematical structures and these laws are somehow unified and symmetrical

+ Our minds are capable of understanding the world and its structure, of seeing the underlying logic of a phenomenon instead of simply experiencing phenomena. For instance, we grasp that matter in motion follows certain rules.

Now this bundle of assumptions was not born in a vacuum, nor did it emerge automatically. As we shall see, many influential thought-forms, ancient and modern, contradict several or most of these assumptions. But if any one of these assumptions is rejected, we cannot do science as we know it. The Matrix includes every one of these assumptions and gives a rationale for each.

It is a matter of historical fact that the discoverers and theoreticians who launched the scientific enterprise consciously held to key provisions of the Matrix. Copernicus and Newton, Faraday and Maxwell, Planck and Einstein, Heisenberg and Dirac were each in their own distinctive ways prophets of the Matrix. We shall show here that they saw a clear relationship between their scientific work and the foundational ideas of the Matrix. Moreover, they went beyond simply accepting the tenets of the Matrix as key assumptions for doing science; they also explored the implications of accepting these assumptions. In other words, each one of them saw a direct connection between the intelligence in the world and an infinite Intelligence.

Certainly there is no evidence that these great scientists were aware of the work of the Four per se, but that's not relevant to our main point. The intellectual formulation of the Matrix probably had a cultural impact that in turn affected the thought-climate of society as a whole, but this is a nebulous connection at best. The crux of our thesis is quite different. *It is that modern science cannot work without conceptual foundations and that these foundations were crafted and established most clearly and comprehensively in the Matrix.* Certainly scientists can do their work without being aware of the foundations of science. Still less do they need to know who provided the rationale for these foundations. But without these foundations they couldn't do science. And when these foundations are abandoned or assailed, the consequences are demonstrably detrimental to science.

To avoid ambiguity, let me specify what is and is not being claimed here:

♦   Central components of the Matrix, e.g., the rationality of the world, were accepted on faith before and after the Four created an intellectual infrastructure to support them.

♦   These central components were required for the birth of modern science.

♦   The creators of modern science affirmed these central components without being influenced in any way by the works of the Four; in fact, the scientists may not even have heard of one or more of them.

But the Four formulated the intellectual rationale for these central components. The Matrix as we refer to it here comprises not simply the components but also the rationale behind them.

In conclusion, the birth of modern science was possible only because its creators assumed that certain beliefs were true. The Matrix as articulated by the Four tells us *why* these beliefs are true. From this standpoint, the Matrix forms a superstructure that supports the foundations of science.

The Big Idea as a Glimpse of the Obvious

Well the Big Idea is in many respects a glimpse of the obvious, and this is the best compliment you can pay it. In actuality the primary truth that energized the Four is the mere fact that things exist; everything else, they said, simply followed from this most obvious of truths. So why is there a need for anything more? A survey of influential ideas past and present shows that many of these emphatically reject the obvious truths that serve as the starting-points of the Matrix. The denial of these truths is fatal for both science and sanity. Hence the urgency of a rational defense and a credible foundation.

Religion as the Enemy of Science

This is an old misconception that simply refuses to die. We have said that the great minds that gave us modern science held a religious view of the world; this religious view was indelibly integrated with their scientific work. Stanley Jaki, the great historian of science, has shown in numerous scholarly volumes, most notably in his Gifford Lectures, *The Road of Science and the Ways to God*, that a belief in the divine creation out of nothing of the universe was essential to the genesis of modern science. Here we will show that there are two major visions of the world, theism, the belief in an infinitely intelligent personal God who created all things, and monism, the belief that all things are essentially just one thing, either a single impersonal Spirit or mere mindless matter. Historically it was theism, the vision supported by the Matrix, which served as the mother of science while monism stood in its way. We have pointed out earlier that the notion of fundamental laws of nature is part and parcel of the theistic vision. As a matter of fact, it is only the rock hard certainty of the religious view of things that sustains scientific belief in the law and order of nature.

Isn't Philosophy a Hindrance to Science?

It's often said philosophy, especially Aristotelian philosophy, obstructed and harmed science. This has certainly been true. But it's not simply Aristotle's philosophy that had this impact. Other philosophies, such as monism, have had the same effect. Nevertheless, science still needs a philosophical framework on which to operate. While adapting what was useful in Aristotle's thought, Avicenna, Maimonides and Aquinas created a dynamic new synthesis that was open to science because, along with Madhvacharya and unlike Aristotle, they recognized an absolute distinction between God and the world and also accepted laws of nature ultimately instituted by God.

Hinduism vs. the Matrix

The idea that Hinduism in its entirety is a form of monism, and therefore opposed to the Matrix, is another popular misconception. Certainly many prominent monists have been Hindus. But only one of the six major schools of Hinduism has promoted monism, and in fact only a subset of this one school. The Hindu thinker Madhvacharya, featured as one of the Four, is, in my view, the single greatest critic of monism in history and a powerful defender of the theistic vision. Madhvacharya's arguments will be presently laid out.

Eastern Mysticism and Modern Physics

In *The Tao of Physics,* Fritjof Capra makes the claim that modern physics has taken us to a view of the world that's very similar to the views held in Eastern mysticism. He takes the latter to mean monism. This claim, as we shall see when I discuss Eastern monism, has been criticized on several fronts. But my main point is that theism, incarnated as the Matrix, historically and conceptually lies at the basis of modern physics. While certain modern popular science writers use the language of monism to describe contemporary findings, it can be shown that monism consistently applied will undermine physics as a whole. The Matrix is the only framework that makes sense of both the foundations and the findings of modern physics.

How do you determine who's the "greatest?"

Although Judaism and Christianity have produced many great thinkers, Maimonides and Aquinas can be described as the "greatest" in their respective religions simply in terms of the influence they've had on successive generations. Islam has also produced many other great thinkers but, arguably, none has rivaled Avicenna in output or influence. The three seminal Hindu thinkers were Sankaracharya, Ramanuja and Madhvacharya. Admittedly, Sankaracharya has had the greatest influence in terms of establishing monism as a dominant view. But I consider Madhvacharya to be the greatest thinker in his tradition because of (a) his definitive and comprehensive refutation of Sankaracharya's arguments and (b) his establishment of a school of thought that produced, among others, the two leading Hindu logicians.

Is the Big Idea/the Matrix an all-or-nothing view of the world?

It is. The Matrix makes certain claims about how the world works, e.g. that it's real and rational. Science assumes that these claims are true and gets to work. The Matrix then goes on to show why its claims about the world are true. Now if the Matrix cannot make a convincing case for the truth of its claims, then it cannot hold them to be true. This means, for instance, that we can't believe in the rationality of the world. The scientist then can't consistently claim to know anything about the world or how it works since there's no rationality at any level.

With these preliminary observations, I propose to lay out my case in the following sequence:

1. The One and the Many

Science, philosophy and religion start with the primordial question of whether there is any unity underlying the many things that make up the world. Three answers are possible. The Many is an illusion and only the One exists. The One is an illusion and only the Many exists. The Many constantly depends on the One that brought it into being.

2. Two Warring Visions of the World

Two fundamental visions of the way things are have battled each other for thousands of years. The first view, monism, holds that there is only one kind of thing, whether it be spiritual or material, while, in contrast, the second, theism, holds that the world proceeds from an infinite Intelligence that is a center of consciousness and intelligence. Monism is popular in some eastern philosophies but five out of six schools of Hinduism are not monist. The monist strain in Hinduism came from the thousand-year reign of Buddhism in India. Monists sometimes sound like theists simply because monism is internally incoherent and therefore hard to describe coherently. Nevertheless it is fundamentally opposed to theism because it holds that, appearances notwithstanding, *only* one thing exists, an impersonal, inactive Reality. Theism on the other hand proposes that an infinite center of Intelligence, will and consciousness brought the world into being.

3. The Three Foundations of Science

Theism has had a distinguished ancestry in various cultures. The Golden Age for theism and human thought was the period when the four greatest thinkers of four world religions crafted an intellectual infrastructure for our study of the world. Key ideas in this structure are identical with the foundations of modern science. The founders of science have tied their scientific foundations to a religious view of the world.

4. The Four Masters of the Matrix

An exploration of common themes lies in the work of Avicenna, Maimonides, Aquinas and Madhvacharya.

5. An Eight-Fold Path to a Theory of Everything

The Matrix provides us with a theistic world-model that's not simply compatible with modern science but makes sense of its discoveries and data. If the success of science is a fruit of the theistic matrix, the next step is to take advantage of all the resources of this matrix.

Chapter

The One and the Many

ROM THE EARLIEST DAYS of human thought, we have been puzzled by the question of the One and the Many. Our everyday experience brings to light a plethora of different kinds of things. Some things are similar to others, different kinds of horses, for instance. Most follow certain common rules of behavior, e.g., gravity. So we ask: Is there some underlying unity that relates things to each other? Are there fundamentally different kinds of realities in the world or is everything made of the same stuff? How is change possible? How did the multiplicity of beings come to be? Is there some underlying source of all things, what has been called the One?

The One has been given different names: the Absolute, God, Saguna Brahman, the Tao and the Void. These questions of unity and synthesis, origin and explanation, are the concerns that drive science, philosophy and religion. The answers are crucial at all levels. Because of their ripple effects, mistakes are potentially catastrophic.

Each discipline gives different kinds of answers to these questions. Science tries to tie things down in terms of the fewest possible laws and aspires towards a single Theory of Everything. Philosophy attempts to go beyond science by proposing systems that unearth meta-scientific relationships and realities. Religion professes to answer questions about the One and the Many by reference to an Ultimate Reality that is the source of all things.

The answers given by the philosophical and religious thinkers are very important for science because they provide it with its meta-scientific foundations. Weak foundations bring down buildings, assuming you can get started at all. An example of a meta-scientific foundation is the assumption that the universe is rational to the extent that it follows certain laws. This assumption cannot be proven by science. But if we don't believe it then we can't do science.

With respect to the One and the Many, the following views have been seriously put forward:

1. The Many is an illusion and the only thing that exists is the One. In other words, there's no "real world;" what we call the world is a misleading veil that hides an Ultimate Reality with which we are identical. The idea that we're distinct from each other is part of the illusion. There's no distinction of subject (one who thinks) and object (what is thought about).

2. The idea of a spiritual One is an illusion. There is only the Many. And the Many is actually matter manifesting itself in different forms, a material One. There are no laws of nature. There's no connection between things such as cause and effect. There's no such thing as mind distinct from matter. At the end of the day, there are only mindless particles and forces. The ideas that we are conscious or have a self are illusions or momentary epiphenomena, not different in kind from matter.

3. We cannot know anything because there is no such thing as truth out there. Our minds are simply brain cells that appear to make associations.

4. The existence of the Many can only be explained by the fact that there is a One that brought it into being. This One has no limitation and has always existed. Its infinite Intelligence and rationality are reflected in the Many, but the One and the Many are entirely distinct from each other. And, by the way, the world is real and does exist as a reality outside our minds.

5. The human person is a union of body and mind, matter and spirit. The mind is not physical, but works closely with the body. It forms concepts and derives insights. It discovers and invents.

6. The mind is capable of grasping the truth in the sense that it can know something to be the case.

We will list the first three views under the category Monism-Skepticism and the last three under Theism-Rationalism. Although there are other options, these two are the historically predominant visions of the world. They are also the views that have had the greatest impact on science and human knowledge. Consequently, we will consider them in more detail.

Chapter

Two Warring Visions of the World

*T*HESE TWO FORMS OF THOUGHT, Monism-Skepticism and Theism-Rationalism, have battled each other on the stage of the human mind from the times of the ancient Greeks, Hebrews, Indians and Chinese to the present.

Monism

The word "monism" derives from the Greek word "monos" which means "one" or "alone." Virtually all forms of monism say the world is made up of only one kind of thing. Some influential versions say that the world itself is just one thing, and all apparent differences are illusory. This vision of the world comes in two fundamental flavors: spiritualist monism and its evil twin materialist monism. *Spiritualist monism says that the only thing that exists is one impersonal spirit. Materialist monism says that everything is matter.* The spiritualists say both matter and the idea of individual spirits are illusions while the materialists say that consciousness, thought and the individual self, cause-and-effect, order, and purpose are illusions.

Now monism is not a familiar term to most people. So why do we use it? Simply because it's more fundamental than other familiar words like atheism. We know what atheism means—it's the belief that there's no God. But in saying there's no Creator or Source of the world, the atheist is also declaring that the world is purely made of matter. This is materialist monism. Thus monism is the ideology that lies beneath atheism, or, to put it another way, atheism is a form of monism.

Monism is also a profoundly skeptical view of the world. Skepticism is the view that no true beliefs about anything are possible; in particular, skepticism denies the possibility of a transcendent being like God or any kind of supernatural activity. The spiritualists say that our minds mislead us into thinking that both the world and we exist. Materialists claim that there's no such thing as the mind; neither is there any order and rationality in the world that would enable us to learn more about things. In different ways, both flavors of monism have led to solipsism, the idea that nothing else exists but oneself.

Theism

Theism derives from the Greek term "theos" for God. Theism affirms the existence of the world and its rationality and the reality of self-consciousness and conceptual thought.

It claims that these hard facts can only be explained if there is a Source of all existent beings that exists without any limitation and is Personal in the sense of possessing intelligence and will.

Since this source, known variously as Ultimate Reality, the Absolute and God, is necessarily free of all possible limitations, it is:

- ◆ omniscient (knows all things),
- ◆ omnipotent (all-powerful),
- ◆ eternal (exists without beginning or end and is outside time),
- ◆ omnipresent (present everywhere),
- ◆ perfectly free,
- ◆ supremely good,
- ◆ immutable (unchanging).

This Being is transcendent, in the sense that it is entirely distinct and infinitely different from the world, but also immanent, in the sense that it keeps the world in being and is present everywhere. The essential insight is that if anything at all exists, then there must exist a source that exists without any limitation and cannot not-exist.

Theism takes the facts of experience on their own terms, accepting the reality of the world and of consciousness and thought while embracing rationality at all levels. Theism addresses the question of why something exists, why there are regularities in the world, why explanations work and how we are able to know things. Theism has been called the ultimate rationalism because it believes both that there is an explanation for everything and that human beings have rational minds that can detect and grasp these explanations. While admitting that we can be in error, theism rejects skepticism and affirms that seeking and finding truth is the normal habit of the human mind. Theism has been championed across the world and throughout history, from the most primeval cultures to Moism in China, Hinduism in India and the thought-leaders of Judaism, Christianity and Islam.

According to theists, monism is wrong because it is built on a denial of what experience and rationality tell us. It is also detrimental. Spiritualist monism makes the Absolute the source of evil, removes all basis for ethics and social action, and makes science impossible because it denies the reality of the world. Materialist monism, with its skeptical view of the world and rationality and its rejection of the human person as a knowing subject, also erodes the foundations of ethics and science.

We will now briefly review the most influential forms of monism in the world, following which we will consider the theist view. At first blush we might wonder why spiritualist and materialist monisms should be lumped together. The answer is that both claim to unify reality by reducing it to one substance, and both distrust everyday experience and common sense. The two kinds of monism have sprung up in both East and West from ancient times to the present.

છ૭

Spiritualist Monism

Although spiritualist monism reached its zenith in the East, it has attracted quite a following in the West as well. Parmenides of Elea (c. 480 B.C.), the leader of the Eleatics, was one of the earliest European monists. He distrusted the evidence of the senses, held that change and plurality were illusions, and maintained that Reality was one. Other famous European monists were the philosophers Baruch Spinoza (1632–1677) and Georg W.F. Hegel (1770–1831). Hegel held that the ultimate and only reality was Mind and history reaches fulfillment when we realize that we are all part of this Mind. Ironically, Karl Marx transformed Hegel's spiritualist monism into a materialist version. The most recent exponent of the spiritualist view in Europe was F.H. Bradley (1846–1924). In his *Appearance and Reality*, Bradley held that ordinary things in our experience are appearances and the only reality is a cosmic experience called the Absolute of which we are a part. There is no distinction between the subject who knows and an object that is known. The only reality is an underlying abstract impersonal consciousness.

The main works of spiritualist monism, of course, came from the East. Most people associate monism with Hinduism and Hinduism with monism. But this habit of thought is simply mistaken. The Hindu scriptures, says the great Hindu scholar B.N.K. Sharma, are largely theistic. He points to "the buoyant realism of the Vedas, the transcendental Theism of the Upanishads and the emotional Theism of the Epics and Puranas." Just as important, monism was taught by only a single sub-set of one of the six schools of Hindu philosophy, a sub-set that chronologically appeared only after the first five schools. The Nyaya, Vaisesika, Sankya, Yoga and Mimamsa schools taught that there is a multiplicity of souls and that these souls are distinct from each other and the world. The Nyaya, Vaisesika and Yoga thinkers specifically contended that God was the intelligent cause of the world. As for the sixth school, that of Vedanta, two of its three distinct versions, the systems of Ramanuja and Madhvacharya, were consciously theistic. Monism as a distinct school of thought in fact appeared relatively late, only in the eighth century A.D.

So how and why did monism appear in Hinduism? The answer is fairly straightforward. Hindu monism was not a fruit of mystical experience or philosophical argumentation. Rather, it was a Buddhist-influenced misinterpretation of the *Vedas* formulated by Sankaracharya as the philosophy of *Advaita* (monist) Vedanta. Gifted followers subsequently popularized it. Ironically, the roots of this supposedly Vedic philosophy can be traced to Buddhism, a movement of thought that had broken off from Hinduism because it rejected the authority of the Vedas! Given the little known fact that Buddhism was the dominant religion in India for over a thousand years, it was all but inevitable that it would influence Hindu thought of the time in some fashion.

From the 2nd century A.D., an influential school of Buddhism argued that the human mind cannot know anything and that the universe is an illusion. Many Hindu thinkers who became Buddhists adopted this idea that the world is an illusion (called *mayavada*). Other Hindus such as Sankaracharya (8th century) adapted Hinduism to include this and related ideas. Sankaracharya held that the only reality in the world is *Niguna Brahman*, an impersonal Absolute with no attributes. Everything else, including the world and our sense of an individual identity, is an illusion and enlightenment comes from realizing our identity with Brahman. From the time of Sankaracharya himself, Hindu critics charged that his monist (*advaita*) version of Vedanta was essentially Buddhist in thrust and content.

We will consider the monisms of the East at greater length in our review of arguments for God's existence.

Materialist Monism

Like its spiritualist counterpart, materialist monism is not a recent phenomenon. The Ionians of Thales (6th century B.C.), Leucippus and Democritus of Abdera (5th century B.C.), Epicurus (342–270 B.C.) and Lucretius of Rome (99 B.C.) held that the only reality was matter. They denied the existence of any mental reality or deity separate from matter. Democritus, for instance, held that the world was made up entirely of an infinite number of tiny atoms that have been in random motion for all eternity.

Materialism as an influential philosophy did not surface again until Julien de La Mettrie (1709–1751), a French physician, published *L'Homme Machine*, a work in which he claimed that humans were self-moving machines. He was followed by the German Baron Paul d'Holbach (1732-1789) who claimed in his *Système de la Nature* that matter in motion is all there ever was and will be. The main materialistic philosophies of the 20th century were Behaviorism and Central State Physicalism, the claim that all mental states are states of the central nervous system.

Since the truth of materialism will be considered at some length elsewhere, here we will simply review its role as a philosophy of monism.

Materialism is not simply the denial of soul or spirit but the much larger claim that matter is the only thing that exists. Matter is understood as a substance that has physical properties, properties that can be physically observed or quantified, such as mass, velocity, position, and charge. At different periods in the history of science, matter has been thought of as particles and forces, fields, strings and branes. But, in all these formulations, there is no doubt that we are talking of physical realities.

Materialist monism claims that there is no such thing as consciousness, intelligence, purpose, freedom, beauty or even laws of nature, although material things may follow certain regular patterns. There are no entities such as God or spirit. There's no self or subject that is conscious, for that matter. Softer forms of materialism don't deny our awareness of being conscious but claim that this is simply an epiphenomenon of matter, another way in which matter manifests itself. There's no rationality of any kind in reality since it's simply incoherent to talk of matter being rational. Anything that happens is a purely random event.

Like spiritualists, the materialists have a skeptical attitude toward truth and knowledge. All acts of thinking are transactions of physical fields. There is no independent entity, the self, which processes the data. While we might describe neural activities as thinking, there is no sensible way in which we can talk of "grasping the truth" or "knowing what is the case" when we're talking simply of fields interacting with other fields.

Although materialists admit that change takes place, these are simply random changes in the states and relations of matter. Every event is entirely determined by prior physical processes.

Materialism received much of its prestige from the success of science. But it is neither the progenitor nor a product of science. Materialist philosophies existed long before modern science got started. But, as we will see, the foundations of science are incompatible with materialism. In its methodologies, science adopts some assumptions also held by materialists. This is because science, as such, studies classes of events and entities that are material. Since it studies physical processes, its methods must be geared toward the observation and measurement of that which is physical. *To suggest that science is materialist because its methods mirror some materialist assumptions is like saying that there is no difference between astronomers and astrologers since they both study the positions of stars.* Whereas the astronomer is simply engaged in the study of stars using tools and methods that will yield accurate data, the astrologer uses the data thus generated to prop up a fanciful philosophy.

Chapter

The Three Foundations of Modern Science

*B*UT MONISM WAS NOT THE ONLY LANGUAGE-GAME IN TOWN. Theism, belief in a personal God separate from the world, one might say, was the normal belief-system of humanity.

Historians of religion like Andrew Lang and Wilhelm Schmidt have argued that primitive religion originated from belief in a Supreme Being. In his classic 13-volume study, *The Origin and Growth of Religion*, Wilhelm Schmidt tried to show that primordial cultures across the world, from Native Americans to Australian aborigines to African tribes, have believed in a supreme God. This God (the Sky-God), sometimes called Father, Creator or simply "Eternal," was worshipped as the creator of all things, beyond all images, eternal, all-knowing, all-powerful and the source of all goodness and of the moral law.

In recent times, scholars have reservations about such hypotheses. Mircea Eliade, for instance, held that it's impossible to precisely identify the point of origin of primitive religion. Eliade saw religion as that which gives unity both to the universe and to our lives. The value of Schmidt's work lies in its refutation of popular reductionist accounts of the origin of religion. Schmidt showed that primitive religion could not have arisen, as previously held, from nature-myths, fetishisms, ghost-worship, animism, totemism and magic for two reasons: (1) These elements could only produce religion after a long process of evolution, and (2) they are not found in the earliest peoples except in certain cases in a feeble form. Schmidt also showed that belief in a "High God" rose very early in human history.

The greatest of the Greek thinkers, Plato (429–348 B.C.) and Aristotle (384–322 B.C.), rejected materialist and spiritualist monism. Both believed in a divinity distinct from the world. Plato's vision of God was of a divine craftsman. Aristotle believed in a God who was Perfect Mind (*nous*), immaterial, eternal and good. Theism in India has its roots in both the Hindu scriptures and in schools of Hindu philosophy. In China, Mo Tzu (c. 479–381 B.C.) was the great voice of theism. Moism espoused the power of human reason and argued from human experience to the existence of God. God, said the Moists, had a loving purpose for the human race. They strongly criticized fatalism while upholding human freedom and the obligation of righteousness and universal love. Judaism, Christianity and Islam each produced great defenders of theism from their very inception.

But all of these developments were simply precursors to the sudden appearance of theism as a comprehensive system of thought, one that would serve as an engine of philosophical, scientific and religious innovation and insight. This emergence of theism as an intellectual superstructure took place over a mere 300+ years. And it was a renaissance that spread its wings from East to West, sweeping before it the suffocating swamps of monism and skepticism. The discovery of the divine went hand in hand with the discovery of the world. In fact, *reflection on the wonder of the world led to a vision of the mind of God that was followed by the birth of modern science.*

The years 980 to 1317 comprised the Golden Age of human thought, an era when the full potentiality of the mind was truly realized and the Matrix of modern science was forged from the ground up.

Between them, Avicenna, Maimonides, Aquinas and Madhvacharya, each one the most notable thinker in his religious tradition, systematically articulated the canons of common sense and everyday experience. They were in a definite sense the *vox populi*, proclaiming simply what was self-evident to all and sundry. They were just as forceful, Madhvacharya most of all, in pointing out that the monist emperors wore no rational clothing.

The Matrix as systematically formulated by the Four was not known to the progenitors of modern science. But the essential truths established and given a rational foundation by the Matrix *independently* played a central role in their scientific work. That is to say, the principles laid out in the Matrix were also the starting-points of modern science.

The central themes in the thought of the Four are outlined below (their actual statements are found in the next chapter). These themes will be studied with reference to the foundations of science.

Principles of the Matrix

Reason—Each of the Four was a devout believer in divine revelation. Equally, each believed in the rightful domain of reason. Precisely because they believed that the human mind was a reflection of the mind of God, they asserted its majesty and reach while granting its weaknesses, so much so that each was criticized by anti-intellectuals in their traditions for their rationalism. What the critics overlooked was the fact that the Four Masters' faith in God justified and demanded their faith in reason. Ironically, history shows that those who lose faith in God soon lose faith in reason as well.

Truth—All four vigorously affirmed the mind's capacity to know what is true. Aquinas and Madhvacharya, in particular, built sophisticated systems defending the mind's ability to grasp truth while also providing classic refutations of skepticism. Since assumptions about the mind's powers are the starting-point of science, these contributions are of enduring importance.

The reality of the world—That things exist was the starting point of the Four Masters' thought and their theism. What does it mean to exist? If something truly exists, what does this tell you about reality? To begin with, these thinkers affirmed and defended the obvious truth that things really do exist, a truth that strangely enough was denied by eminent philosophers. From this great truth, the Four were led to the other key principles: the world as rational; the self; and God.

The reality of the self—Many influential thinkers and thought-forms start with the assumption that the self does not exist. Materialists and spiritualists revel in the revelation of their non-existence as subjects or centers of thought and consciousness. But such intellectual irresponsibility was refuted and repudiated by the Four. It is not too much to say that the human race owes them a great debt for their painstaking analysis of the structure of the self, of our capabilities of thinking, willing, perceiving, feeling, imagining, dreaming and communicating,

The existence of God—The center of their thought was, of course, God. But it was an understanding of God that started with the world, supplemented to be sure with the insights of their scriptures. As they saw it, the world did not make sense on its own. Since they held that ultimately things do make sense, they embarked on a journey that started with the intelligence and energy and structure in the world and ended in an ever-existent, infinitely intelligent, all-powerful source. Paradoxically, the discovery of the divine source lent a new impetus and confidence to the exploration of a world that was now rationally grounded.

The relationship of God and the world—This was THE issue for the spiritualists. There was no way in which they could reconcile the existence of anything separate from the One. But for the theist, there was no conceivable way in which the world with its imperfections could be seen as part of a transcendent infinitely perfect being. Equally, there was no way to deny the reality of either the world or oneself. The answer lay in drawing attention to the total and continuing dependence of the finite on the Infinite. While there could be no compromise on the absolute distinction between God and the world, the very insight that pointed to the infinite gulf between them also affirmed their infinite intimacy.

The nature of God—Certainly the Four based some part of their understanding of the nature of God on the religious writings of their traditions. But they were quite aware of the need to draw on the resources of reason and experience in exploring the divine attributes. As a result, they constructed rational models for thinking about God. All four held that God, being a center of intelligence and will, was personal. They showed also that God was necessarily without limitation. Starting with this assumption, they held that God was One, eternal and perfect.

Monists and Atheists—None of the Four lived or thought in a vacuum. The same skepticism that is so common today was alive and well in their intellectual environments. Maimonides' major work was addressed to a disciple who was "perplexed" by the doubts sown by philosophers. Aquinas structured his entire *Summa Theologica* as responses to specific objections. Madhvacharya actively debated monists and skeptics all across India. It is not too much to say that most possible objections to the existence of God, the soul and truth have already been addressed in the writings of the Four Masters.

Now what, you might ask, does all this have to do with the foundations of science? It's easy to see that some sort of belief in reason is required to do science, but not at all easy to see why belief in God should make any difference. In fact, some have said that it's a positive obstacle to the scientific enterprise. I propose to address this common misconception in two stages. First, I hope to show that the principal assumptions of the theory and practice of science require a theistic basis, such as is laid out in the Matrix. Second, the scientists who founded modern science bear witness to the root-to-fruit symbiosis of religion and science. Their comments will show that these scientists are the Prophets of the Matrix.

The Three Foundations of Modern Science

It is a matter of historic fact that modern science began in theistic cultures infused by theistic visions of the world. But just as important, the foundations of science are inextricably embedded in a theistic view of things.

I think no one would disagree that these three working hypotheses are taken for granted by science. They are the three foundations of science:

1. The world truly exists and is different from myself. It is not a deity or part of a deity.

2. The things of the world behave in accordance with certain regularities and all the phenomena in the world can be explained in some fashion or other, even if these phenomena involve explanations of a different kind at the quantum level.

3. I exist and can know things about the world. In other words there is an "I" who can perceive things and also think and grasp what is the case. I can observe phenomena, discern connections between different phenomena, arrive at certain insights about these phenomena, capture these insights in concepts that have a specific meaning for other "I"s like myself, string these concepts into a theory that seems to explain these phenomena, and finally, communicate the theory to a community of others.

The hypotheses underlying the scientific enterprise as well as their successful application in discoveries and inventions tell us certain things about the world:

◆ Reality as a whole is rational, and so it can be understood, categorized and connected using rational principles.

◆ There's a relationship of cause and effect, of events and explanations, that applies everywhere in the world.

◆ There's an underlying unity, symmetry and simplicity in the laws that govern the world, and scientific theory is therefore a process of progressive unification. Science has shown levels of interconnectedness, for instance at the levels of both the microverse and the macroverse, that we would never have imagined if we didn't have the tools and methodologies, frameworks and perspectives of present-day physics.

◆ There has to be a certain degree of stability and regularity if scientific theory and investigation is to be possible.

◆ Science assumes and finds intelligence in both animate and inanimate systems. This ranges from real-time information processing to autonomous agency.

◆ Not only can the world be known, we can know it. This mysterious correspondence between mind and world is taken for granted, but not by a reflective mind. Einstein said that the most incomprehensible thing about the world is the fact that it's comprehensible.

This scientific vision of the world just happens to be identical with the world-vision of the Matrix. Both appeal to the undeniable reality of ultimate explanation, intelligibility, rationality, coherence and unification. It might be asked if any discovery has cast doubt on any of these principles. I reply that, on the contrary, every discovery consolidates and amplifies our awareness of the intelligence at work in the world. We note also that the great theories of science are popular because of their explanatory power, because they make sense of things. They confirm our conviction that there is an explanation for everything and that we can find it.

Now the initial three hypotheses, the foundations of science, cannot themselves be proved by science.

1. If the world were said to be an illusion, science by itself would not be able to show that this is false; any attempt to demonstrate the reality of the world could be dismissed as part of the illusion. Likewise, it could not disprove the claim that the world and we are part of a deity.

2. Similarly, the fact that processes seem to follow certain laws or that explanations are obtained if we do experiments does not prove that there is regularity and cause-and-effect in the world. Quite conceivably, these observed behaviors could be sheer coincidences and reality could actually be random; in a truly random world there can be no laws and theories.

3. Finally, science does not have the equipment to show that the knowing self exists or that this self is capable of rational thought and valid arguments. Neither the self nor its capacity to know can be physically observed or measured.

The Nobel Laureate Charles H. Townes draws pointed attention to the fact that science is based on faith in scientifically unprovable foundations:

> "We approach science with a kind of faith. It is a faith in our logic and ability to understand...We have to accept a certain framework in which to operate. We scientists believe in the existence of the external world and the validity of our logic. We feel quite comfortable about it. Nevertheless, these are acts of faith. We can't prove them.
>
> [Belief in the existence of the external world] is part of the faith of science. Faith is, in part, believing that the things I see around me are not just my imagination. They are real. I can do experiments, and I may still be fooled. But every scientist now accepts the reality of a universe governed by physical laws; it is so deeply embedded that we never think of it as faith. And yet that is just what it is."

But without faith in these foundations, it is all but impossible to get science on the road:

♦   If we start off with the belief that the world is an illusion or that the only thing that exists is myself or that the world is part of a God, then we have neither motivation nor rationale for doing science.

♦ If we seriously think that everything is random, why on earth should we even try to find laws or concoct theories?

♦ If we are nothing but an assemblage of atoms, then we should have no confidence in our judgments and, ipso facto, we shouldn't waste time collecting facts and supposedly connecting the dots.

Psychologically and logically, there would be no basis for science. It may be said that we can mentally cordon off our beliefs about the world and create a separate compartment for doing science. But we can't fool even ourselves all of the time. Beliefs and practices inevitably influence each other.

The three foundations are not simply hypotheses but the very conditions required for any possibility of science. It's not surprising, then, that the idea of fundamental laws of nature, the backbone of modern science, arose in theistic cultures. *And it's a striking fact of history that Copernicus, Galileo and Kepler, Newton, Faraday and Maxwell, Einstein, Planck and Heisenberg, all believed in a divine Mind behind the world and Rationality at the foundation of reality.* To this we shall now turn.

Prophets of the Matrix

At the heart of the Matrix is the affirmation that the rationality of the universe can only be explained by the existence of infinite Intelligence, an omnipotent, omniscient God. This is its most controversial claim, but it's also the wellspring of all its other principles. Oddly enough, this was the very claim that was ardently and unmistakably affirmed by the scientists who generated modern science.

The reflections of scientists like Copernicus, Kepler, Galileo, Newton, Maxwell, Planck, Einstein, Heisenberg, Dirac and Born on rationality and God and on the common source of science and religion are profoundly insightful, inspirational and touching. In depth and power, they mirror the insights of the Matrix.

We have already seen how the foundations of science are inevitably embedded in a theist-rationalist vision of the world, the Matrix. But nothing brings this relationship home to us as vividly as the testimony of the great scientists themselves as seen below.

Nicolaus Copernicus, Heliocentric Theory of the Solar System

♦ "How exceedingly vast is the godlike work of the Best and Greatest Artist!"

♦ "The Universe has been wrought for us by a supremely good and orderly Creator."

Johannes Kepler, Kepler's Laws of Planetary Motion

♦ "Praise and glorify with me the wisdom and greatness of the Creator, which I have revealed in a deeper explication of the form of the universe, in an investigation of the causes, and in my detection of the deceptiveness of sight."

♦ "[May] God who is the most admirable in his works...deign to grant us the grace to bring to light and illuminate the profundity of his wisdom in the visible (and accordingly intelligible) creation of this world."

♦ "The creator chose nothing without a plan."

Galileo Galilei, Laws of Dynamics, astronomical confirmation of the heliocentric system

"The holy Bible and the phenomena of nature proceed alike from the divine Word."

Isaac Newton, Optics, Laws of Motion, Gravitation

Newton's theological writings, running into a million words, far exceeded his scientific output. Below is an excerpt from his classic work, the *Principia Mathematica*:

"This most beautiful system of the sun, planets and comets could only proceed from the counsel and dominion of an intelligent and powerful Being. This Being governs all things not as the soul of the world, but as Lord over all; and on account of his dominion he is wont to be called 'Lord God'…or 'Universal Ruler.'…And from his true dominion it follows that the true God is a living, intelligent and powerful Being…he governs all things, and knows all things that are or can be done. He endures forever, and is everywhere present…Blind metaphysical necessity, which is certainly the same always and everywhere, could produce no variety of things. All that diversity of natural things which we find suited to different times and places could arise from nothing but the ideas and will of a Being necessarily existing."

James Clerk Maxwell, Electromagnetism, Maxwell's Equations

♦ "One of the severest tests of a scientific mind is to discern the limits of the legitimate application of the scientific method."

♦ "Science is incompetent to reason upon the creation of matter itself out of nothing. We have reached the utmost limit of our thinking faculties when we have admitted that because matter cannot be eternal and self-existent it must have been created."

♦ "I have looked into most philosophical systems and I have seen that none will work without God."

Lord William Kelvin, Laws of Thermodynamics, absolute temperature scale

"I believe that the more thoroughly science is studied, the further does it take us from anything comparable to atheism."

Albert Einstein, Theories of Relativity

♦ "I have never found a better expression than 'religious' for this trust in the rational nature of reality of reality and of its peculiar accessibility to the human mind. Where this trust is lacking science degenerates into an uninspired procedure. Let the devil care if the priests make capital out of this. There is no remedy for that."

♦ "Whoever has undergone the intense experience of successful advances in this domain [science] is moved by profound reverence for the rationality made manifest in existence...the grandeur of reason incarnate in existence."

♦ "Certain it is that a conviction, akin to religious feeling, of the rationality or intelligibility of the world lies behind all scientific work of a higher order...This firm belief, a belief bound up with deep feeling, in a superior mind that reveals itself in the world of experience, represents my conception of God."

♦ "I want to know how God created this world...I want to know His thoughts, the rest are details."

Max Planck, father of Quantum Physics

♦ "There can never be any real opposition between religion and science; for the one is the complement of the other."

♦ "Religion and natural science are fighting a joint battle in an incessant, never relaxing crusade against skepticism and against dogmatism, against unbelief and superstition...[and therefore] 'On to God!'"

J.J. Thompson, discoverer of the electron

"In the distance tower still higher [scientific] peaks which will yield to those who ascend them still wider prospects and deepen the feeling whose truth is emphasized by every advance in science, that great are the works of the Lord."

Werner Heisenberg, quantum physicist, Heisenberg Uncertainty Principle

♦ "In the course of my life I have repeatedly been compelled to ponder on the relationship of these two regions of thought [science and religion], for I have never been able to doubt the reality of that to which they point."

♦ "Wolfgang [Pauli] asked me quite unexpectedly: 'Do you believe in a personal God?'...'May I rephrase your question?' I asked. 'I myself should prefer the following formulation: Can you, or anyone else, reach the central order of things or events, whose existence seems beyond doubt, as directly as you can reach the soul of another human being. I am using the term 'soul' quite deliberately so as not to be misunderstood. If you put your question like that, I would say yes…If the magnetic force that has guided this particular compass—and what else was its source but the central order?—should ever become extinguished, terrible things may happen to mankind, far more terrible even than concentration camps and atom bombs.' "

Arthur Compton, quantum physicist, Compton Effect

"For myself, faith begins with a realization that a supreme intelligence brought the universe into being and created man. It is not difficult for me to have this faith, for it is incontrovertible that where there is a plan there is intelligence—an orderly, unfolding universe testifies to the truth of the most majestic statement ever uttered—'In the beginning God.' "

Max Born, quantum physicist

♦ "Those who say that the study of science makes a man an atheist must be rather silly."

♦ "Something which is against natural laws seems to me rather out of the question because it would be a depressive idea about God. It would make God smaller than he must be assumed. When he stated that these laws hold, then they hold, and he wouldn't make exceptions. This is too human an idea. Humans do such things, but not God."

Paul A.M. Dirac, quantum physicist, matter-anti-matter

"God is a mathematician of a very high order and He used advanced mathematics in constructing the universe."

◈

Einstein and God

It might be asked if Einstein really believed in God. Isn't it well known, in fact, that he actually rejected belief in a personal God?

It is true that on more than one occasion, Einstein said that the idea of a personal God is an anthropomorphic concept that he could not take seriously. At the same time, it is evident from his own comments that he thought of God as a center of consciousness and intelligence. In studying this question, we keep three things in mind. Einstein himself said that the man of science is a poor philosopher and it is unfair to expect from him a fully developed philosophy of God. Second, he was influenced by Spinoza's ideas of an impersonal Absolute. Third, he was rightly concerned about the danger of seeing God in our own image as some kind of super-person and of violating the laws of cause-and-effect.

As we see above, Einstein very specifically talked of God as an intelligent spirit. He famously said that he wanted to know how God created the world; in fact, he wanted to know "His thoughts." He also said "my God" created laws so that the universe is governed by immutable laws and not wishful thinking. None of these comments make any sense if we assume that God is an impersonal force. Einstein also said that it made him angry when people call him an atheist, and once even said belief in a personal God is better than no transcendental outlook of life.

His concerns about the idea of a personal God arose from two factors: the fear that we may create a god in our image and the conviction that there are no exceptions to the law of causality such as would be the case in the event of divine intervention. The first concern is certainly valid, but the way to address this is by enlarging our idea of God, not reducing it. God is radically different from all finite being, although His attributes can be known through analogy. He is wholly transcendent and we should exorcise all anthropomorphic notions. But we should not replace these notions with bizarre ideas of forces that exhibit personal behavior without being personal. As for exceptions to the law of cause and effect, the mistake is to assume that the only things in the world are physical or material. Once we recognize the reality of the mental, we can see that physical laws of cause and effect do not bind it, except when it interacts with the physical.

In his correspondence with Einstein, the theologian Paul Tillich addressed many of these concerns. "The depth of being," wrote Tillich, a fellow ex-ile from Germany, "cannot be symbolized by objects taken from a realm which is lower than the personal, from the realm of things or subpersonal living beings. The suprapersonal is not an 'It,' or more exactly, it is a 'He' as much as it is an 'It,' and it is above both of them. But if the 'He' element is left out, the 'It' element transforms the alleged suprapersonal into a sub-personal, as it usually happens in monism and pantheism." As a footnote, it's worth mentioning that Einstein, when told by G.S. Viereck that his theory of relativity was compatible with the thought of Thomas Aquinas, remarked, "I have not read all the works of Thomas Aquinas, but I am de-lighted if I have reached the same conclusions as the comprehensive mind of that great Catholic scholar."

On balance, it is indisputable that, while denying that God had the attri-butes of human personality, Einstein accepted a transcendent Mind as the source of the rationality of the world.

The Curious Case of Charles Darwin

The preceding catalog of convictions raises an obvious question: what about Darwin?

The case of Charles Darwin is quite complex. At one time Darwin was clearly a theist, and in his public writings he talked of "laws imprinted on matter by the Creator." But in later life, he became an agnostic and pos-sibly an atheist, writing that "the whole subject is too profound for the human intellect;" some have said that the premature death of his daugh-ter had a devastating effect on his early faith. His intellectual environ-ment, in particular William Paley's (1743–1805) version of natural theology, shaped Darwin's ideas about God. Ironically, he eventually considered Paley his nemesis. Moreover, Paley's Maker analogy (see below) still gov-erns the current evolution-creationism debates. But the Paley approach has nothing to do with the essential thrust of the Matrix. Besides, neither Darwin nor the pre-eminent evolutionary theorists seem to have had any idea of what was proposed and systematized in this meta-scientific Theory of Everything.

Ever since Robert Boyle (1627–1691), noted theists in England began to present the case for God within the framework of natural science. They operated within a mechanist view because they saw all things as components of a world machine. The mechanisms of the world that could not be scientifically explained, e.g. certain kinds of motion, were attributed to the direct action of God and cited as evidence of His design. This approach, often called physico-theology, had several shortcomings:

♦ The domains of the theologian and the scientist overlapped to such an extent that new discoveries in science progressively eroded what was left of theology. Mechanisms that were thought to be of supernatural origin were found to have natural causes.

♦ The method of argument was strictly one of analogy. William Paley, for instance, noted that the discovery of a watch on a moor would suggest the existence of a watchmaker. Likewise, he argued, the intricate mechanisms in nature can only be explained by the existence of a Maker. The strength of the argument lay in the strength of the analogy.

♦ The theory did not focus on the fact that things exist and therefore required a ground of their existence. Nor was there any grasp of the need for ultimate causes and explanations.

These shortcomings were devastating for physico-theology. Discoveries in celestial mechanics and geology had radically undermined its credibility. Paley sought to make a last stand in biology by pointing to the adaptations of organisms in nature and the impressive functions performed by each one of their components. This seemed to give clear evidence of direct divine design and action. Darwin, ironically, had once accepted the validity of this argument. But then he proposed a natural mechanism for achieving precisely the same results that were thought to be of divine origin. According to his theory, the struggle for existence, random variation, domination of advantageous properties, and the survival of the fittest explain all that takes place in the organic order. With one stroke he discredited Paley's analogy and therefore his argument: "The old argument from design in Nature, as given by Paley, which formerly seemed to me so conclusive, fails, now that the law of natural selection has been discovered." And since Darwin's theism was based on physico-theology, the end of physico-theology also spelled the end of his theism.

But Paley was one thing and the Matrix another. Neither Paley nor Darwin came to grips with the central issues laid out by rational theism:

- Why is there something rather than nothing?

- What does it mean to exist?

- How did reality come to be structured such that there are fundamental laws of nature and a hierarchy of intelligence in the natural world? Even if natural selection is cited as the mechanism for this state of affairs, our question remains unanswered since the mechanism can only work within a structure. How did this structure originate?

- What tells organisms to adapt to environment? How did organisms develop the capacity for replication, an intrinsically purposive activity, and one that was essential for the survival of living beings?

- What are life, consciousness, and conceptual thought? How could they spring forth from undifferentiated matter or, if you prefer, an energy field?

These and many like them are meta-scientific questions that fall outside the realm of observation and measurement. But science implicitly assumes that there are answers to these questions that are both rational and plausible. And the only viable answers are those crystallized in the Matrix. As for Darwin, Loren Eisely rightly noted that he destroyed not the design argument, but the watchmaker and the watch. This elimination of the Cosmic Machinist opens the door to the horizon presupposed by science from the beginning: the recognition of the infinite Intelligence that is the Source of existence.

We shall now see how Avicenna, Maimonides, Aquinas and Madhvacharya approached these themes.

Chapter

The Four Masters of the Matrix

BY SOME EXTRAORDINARY COINCIDENCE, the four greatest thinkers of four major world religions, Hinduism, Judaism, Christianity and Islam, lived and worked within a relatively short window of time, from 980 to 1317. We have sketched some of the common themes in their thought.

In *Knowing the Unknowable God,* David Burrell shows that Avicenna, Maimonides and Aquinas, in their own ways, molded Hellenistic and Hebraic thought into a new revelation of God's transcendence. In all three thinkers, we see God presented both as the source of all things, particularly rational creatures, and as entirely distinct from them. God is totally outside the universe but immanent inasmuch as he holds it in being at every instant.

The fourth thinker in this study, Madhvacharya, had no connection with Hellenistic or Hebraic thought. Textually and historically, it is clear that he had no contact with Western or Judeo-Christian-Islamic thinkers and thought-forms. Nevertheless his vision of God and the world mirrored that of the three others. Between them the Four represented the two most ancient and most refined religious world-outlooks in history, the Hebraic and the Hindu.

Here we will briefly consider the life and thought of each one of the Four individually.

Avicenna

Avicenna (980–1037), Abdaallah Ibn Sina, known as "The Supreme Master," was the greatest of the Islamic thinkers. Born in Bukhara, Persia, he became physician and adviser to sultans and princes. His *Canon of Medicine*, written at the age of 21, was the best-known medical text in Europe and Asia for several centuries. He authored over a hundred works in medicine and philosophy that have inspired innumerable commentaries. His most important books in philosophy were *The Healing* (*al-Shifa*) and *Demonstrations and Affirmations.* He died in Hamadan in northern Persia.

Avicenna made enduring contributions to the areas under discussion here. He is especially famous for his insights into the necessary existence of God and the non-material nature of the human soul.

While physics is concerned with the motion of things, metaphysics focuses on the very existence of things. Why and how is it that they happen to exist? There is no scientific or logical law that says they must necessarily exist. There is only a possibility of their existing, and an equally real possibility that they might not have existed. But, unlike all other beings, God exists by necessity, and his non-existence is impossible. To exist belongs to the very essence or nature of God. He exists, and cannot not-exist.

The existence of beings that do not exist by an inner necessity of their natures points to the existence of the necessary being, God. Even an infinite chain of these beings that are each caused to exist by a source external to itself cannot explain how any or all of them came to exist. Only a first cause that exists necessarily can explain the existence of every other being.

"This is what it means that a thing is created, that is, receiving its existence from another," writes Avicenna, "As a result everything, in relation to the first cause, is created...Therefore, every single thing, except the primal One, exists after not having existed with respect to itself." That is, anything brought into existence by the first cause requires the action of this cause to remain in existence. Avicenna writes, "That which is caused requires something which bestows existence upon it continuously, as long as it continues as existing."

No cause is required to explain the existence of a necessarily existing being. Avicenna observes, "That whose existence is necessary through itself does not have a cause while that whose existence is possible through itself does have a cause." And there can be only one necessary being. "That whose existence is necessary must necessarily be one essence" is the first volley of his elaborate argument to prove this particular thesis.

Like other thinkers influenced by Aristotle and Plato, Avicenna maintained that there was a hierarchy of intelligent beings in the universe. This scheme led some critics to call him a pantheist. But these accusations are implausible, given that it was Avicenna who underscored the radical difference between God, the necessary being whose essence is to exist, and all other beings.

Although Avicenna believed that the world is a creation of God, he also believed, under the influence of Aristotle, that both God and the world existed eternally. As Aquinas and other theists acknowledge, this view is not self-contradictory because creation does not necessarily require a beginning in time. Avicenna, of course, noted that in itself the world is only "possible" and requires a cause for its existence. God, on the other hand, exists necessarily and brought the world into being from nothing. This act could either have a beginning or be beginningless and endless. Other Islamic philosophers put forward the *kalam* argument, made famous in recent times by William Lane Craig, which shows that the universe had to have a beginning in time. F.F. Centore observes that one defect of Avicenna's thought was his assumption that the world necessarily emanates from God.

Avicenna also introduced innovative arguments to show that the human soul is immaterial and indivisible. He noted that each person is ineradicably aware of his/her existence as an individual self, a self that will permanently retain its individuality.

☙

Moses Maimonides

Moses Maimonides (1135–1204), Rabbi Moshe ben Maimon (Rambam), was the most influential Jewish thinker since, well, Moses. A Jewish saying makes this very point: "From Moses [of the Torah] to Moses [Maimonides] there was none like Moses." Born in Córdoba, Spain, he fled to Morocco and then settled down in Egypt after the intolerant Almohads gained power. In later life he became the court physician of the Sultan Saladin and the head of the Jewish community in Cairo. Maimonides' most famous work is *Guide to the Perplexed,* an explanation of God's infinite perfection addressed to a disciple who was troubled by disputes in philosophy and theology. He also authored several classic works of Jewish law and scriptural commentary.

From the present standpoint, Maimonides' perspectives on God, the world and human reason are of special interest:

> "As for that which has no cause for its existence, there is only God...His existence is necessary. Accordingly, His existence is identical with his essence, and his true reality, and his essence is his existence."

In his *Commentary to the Mishnah,* he says, "God is the Being, perfect in every possible way, who is the ultimate Cause of all existence. It is inconceivable that God does not exist, for if God did not exist, everything else would also cease to exist and nothing would remain. Only God is totally self-sufficient and, therefore, Unity and Mastery belong only to God."

God, writes Maimonides, "is one in all respects; no multiplicity should be posited in Him...the numerous attributes...that figure in the Scriptures and that are indicative of Him...are mentioned in reference to the multiplicity of His actions and not because of a multiplicity subsisting in His essence." In other words, God is Creator of all things and eternal. In Part 2 of the *Guide,* Maimonides gives 26 propositions that demonstrate the existence of God and his attributes of being One and without a body.

God created the world from nothing. Maimonides admits that creation is not provable from reason. But the eternal existence of the world is similarly unprovable. Creation, however, is more likely because it is compatible with

God's freedom and is a better explanation of how the multiplicity of the world could originate from the unity of the divine.

By showing that the Creator is infinite Intelligence, he provided a solid foundation for our belief that the universe and its laws are rational:

> "We believe that the Universe remains perpetually with the same properties with which the Creator has endowed it, and that none of these will ever be changed except by way of a miracle in some individual instances, although the Creator has the power to change the whole Universe, to annihilate it, or to remove any of its properties. The Universe had, however, a beginning and a commencement, for when nothing was as yet in existence except God, His wisdom decreed that the Universe be brought into existence at a certain time, that it should not be annihilated or changed as regards any of its properties, except in some instances; some of these are known to us, whilst others belong to the future, and are therefore unknown to us."

Unlike pantheists and monists, Maimonides held that God totally transcends the world. This avowal has enormous consequences. While Aristotle's idea of purpose was focused on the purposive structures in the world, Maimonides argued that a true purpose for the universe could only come from something outside it. Since God transcends the universe, he can give an overarching purpose to the universe as a whole. Interestingly, his greatest Jewish antagonist in subsequent centuries was the pantheist Baruch Spinoza.

Like most theists, Maimonides believed that none of our verbal descriptions adequately describe God. He emphasized the negative approach to God's attributes. We speak of them more in terms of what they are not; for instance, God is not limited by time or by space. The attributes of God catalogued in the scriptures, he said, "are mentioned only to direct the mind toward nothing but His perfection [or]...are attributes referring to actions proceeding from Him."

Maimonides had a positive view of the mind's ability to discover truth. He held that there could be no conflict between the truths we discover in science and philosophy and the truths we believe to be revealed from

God. He also did not believe that religious authorities could answer scientific questions. The human intellect reaches the summit of its powers, said Maimonides, when it studies God, who is absolute, pure Intelligence, since the ultimate goal of life is to know God and love him. We enhance our intellectual growth by studying nature and mathematics, and in fact we come to know God better precisely through these endeavors. But to be truly fulfilled we must also aspire to moral perfection.

Thomas Aquinas

St. Thomas Aquinas (1224/5–1274), called the Angelic Doctor, was the foremost Christian philosopher in history. Thomism, the school of thought built around his work, has attracted disciples from both different religions and no religion. Born to noble parents, he became a monk in the Dominican Order in 1243. He studied under Albert Magnus and taught at the University of Paris. Before he died at the age of fifty, he authored numerous works of philosophy and theology that came to some 8 million words. The *Summa Theologica* and the *Summa Contra Gentiles* are his two most celebrated books.

Aquinas made distinctive contributions in multiple areas, but our concern here is with his writings on God, the soul and human reason.

God

Aquinas embraced Avicenna's formulation of God as the Being that exists by its very nature/essence, and refined the formulation further by introducing the idea of act and potency. An act is any activity or operation while potency is a potentiality or capability. A car has the passive potency of running on the road and its driver has the active potency of being able to run it. The actual running of the car is the actualization of both potencies. These ideas are transferred to the question of existence. The existence of a being is the act of existing of its essence, of what it is. All beings other than God are dependent for their existence, for the actualization of their essence, on external, previously existing beings. Ultimately, they are dependent on God because His essence is identical to His act of existing. "Everything, then, which is such that its act of existing is other than its nature," writes Aquinas, "must needs have its act of existing from something else. And since every being which exists through another is reduced, as to its first cause, to one existing in virtue of itself, there must be some being which is the cause of the existing of all things because it itself is the act of existing alone...There is a being, God, whose essence is His very act of existing."

God is Pure Act. He cannot exist more fully than he does and has no passive potentiality yet to be actualized. "The act of existing which is God is such that no addition can be made to it," said Aquinas. "God possesses all perfections in His very act of existing."

Aquinas believed that human reason could discover the existence of God from reflection on the world. From effects we find the cause. This is the basis of his famous Five Ways for demonstrating the existence of God, which are five versions of one fundamental insight: the affirmation that all existent beings must be the ground of their own existence or have this ground in something else.

Although Aquinas believed that our knowledge of God proceeds from negation, from determining what God is not, what limitations he lacks, he also believed that we can know something about God by way of analogy:

> "God prepossesses in Himself all the perfections of creatures, being Himself absolutely and universally perfect. Hence every creature represents Him, and is like Him, so far as it possesses some perfection: yet not so far as to represent Him as something of the same species or genus, but as the excelling source of whose form the effects fall short, although they derive some kind of likeness thereto. ...When we say *God is good*, the meaning is not, *God is the cause of goodness*, or, *God is not evil*; but the meaning is, *Whatever good we attribute to creatures pre-exists in God*, and in a higher way...He causes goodness in things because He is good."

Laws of Nature

> "There is a certain Eternal Law, to wit, Reason, existing in the mind of God and governing the entire universe."

The Soul

> "The nature of anything is manifested from its operation.

Now, the proper operation of man, as man, is to understand; indeed he rises above all else by this operation...The
intellectual principle is the proper form of man."

"We have to say that the principle of intellectual operation
which we call the soul of man, is some sort of incorporeal
and subsistent principle...This intellectual principle
called mind or intellect has an operation by itself which it
does not share with the body."

Truth

"The nature of the true consists in a conformity of thing
and intellect."

"Our knowledge, taking its start from things, proceeds in
this order. First, it begins in sense; second it is completed
in the intellect."

"First principles are immediately known when we know
their terms. ... The intellect is not deceived in any way
with respect to first principles. It is plain, then, that if
intellect is taken in the first sense – according to that action from which it receives the name 'intellect' – falsity
is not in the intellect. Intellect can also be taken in a
second sense – in general, that is, as extending to all its
operations, including opinion and reasoning. In that
case, there is falsity in the intellect. But it never occurs if
a reduction to first principles is made correctly."

Madhvacharya

In terms of personal charisma, intellectual rigor and scholarly breadth and
depth, Madhvacharya (c. 1238–1317) was the most fascinating of the Hindu
sage-philosophers and one of the greatest theistic thinkers of all time. He
was an Indian Wittgenstein whose rapier-sharp critiques matched his
memorable and profound aphorisms. More to the point, he was an intellectual juggernaut who single-handedly reversed the slide toward monism
and re-established theism as a dominant force. He was also an accom-
plished wrestler, mountaineer and singer!

Born near Udupi in South India, he left his family at the age of 16 (some accounts say 12) to take up life as a religious ascetic. As was common in those times, he had a guru (teacher) who was responsible for his intellectual and spiritual formation. The guru, like most of his contemporaries, was under the spell of *Advaita* (monist) Vedanta. But from the beginning, Madhvacharya would trust only his own experience and the principles of reason. Rejecting *Advaita* on rational and religious grounds, he systematically laid out the case for theism, eventually convincing even his guru. He visited the major intellectual centers of the day, debating monists and drawing attention to the theism of the Hindu scriptures. By the time of his death he had written 37 books, converted the most prominent *Advaita* scholars in India to theism, and assembled eight disciples to carry on his work. His defense of theism and his critique of monism were continued by numerous subsequent thinkers, most notably Jayatirtha and Vyasatirtha, the two greatest logicians in the history of Indian thought.

Madhvacharya's task was two-fold: (1) to show that theism is taught by experience, reason and the Hindu scriptures and (2) to refute the monism that was popular in his time. He was motivated by four principles:

1. A determination to remain true to experience above all, in the spirit of science

2. A commitment to sound reasoning

3. A fervent devotion to a personal God that drove all his actions

4. Fearless tenacity in expounding his vision in the most hostile environments

The underlying theme in all Madhvacharya's work was his famous exposition of the five differences:

> "The difference between the *jiva* (soul) and *Îshvara* (Creator), and the difference between *jada* (insentient things, e.g., matter) and *Îshvara*; and the difference between various *jîvas*, and the difference between *jada* and *jîva*; and the difference between various *jadas*, these five differences make up the universe." —*Madhvacharya (quoting from the* Paramopanishad *in a commentary on the Hindu scriptures)*

Madhvacharya presented a very simple vision of the world. It was clear to him that there were differences and distinctions in the world. Matter was distinct from mind. One material thing was distinct from another, one person from another. Above all, there was a radical difference between God and the world. This in a nutshell is his doctrine of *Panchabeda* or five differences, which stated that there was an absolute distinction between God and the soul, God and matter, souls and matter, each individual soul and another, and each material thing and another. There is an unbridgeable gulf between God and all other beings because God is the only independent Reality.

The theme of difference, individuality and uniqueness is fundamental in Madhvacharya's thought as it was for John Duns Scotus in the West. By the very fact that something is what it is, says Madhvacharya, it is obviously different from everything else, and this is shown to us by both reason and our senses. The substance of each particular thing is a unique combination of many properties. While many other things could possess these same properties, the difference between each and every thing is the uniqueness of the specific combination of properties. At the very least there is a difference in location for physical things. Scotus spoke of this same uniqueness as "the individualizing" that makes one thing different from another.

And the source of all these properties and their unique combination is God the great inventor and sustainer. "God Himself," said Madhvacharya, "is the determining cause of the distinctive natures of the various tastes, their essences and their characteristics themselves, in a special sense. It is not to be understood that those special characteristics and essences are determined by the intrinsic nature of the substances themselves. Far from it. Not only the substances, but their respective essence and characteristics and the characteristics of those characteristics themselves are *all* derived from his immanent powers and presence in them."

In understanding the five differences, we come to grasp the properties of all the things in the world and the relationships between them. Most important of all, we come to realize our total dependence on God.

Starting with the five principles, Madhva focused his attention on three areas:

1. *How We Know.* We are able to know what is the case about things through three sources: experience, reason and divine revelation. The primary guarantor of truth and certainty in our coming to know something is a capability he called *Sakshi.* His theory of

knowing and truth is very important because it stands in sharp contrast to the skepticism of his contemporaries.

2. *God and the World.* Reality may be divided into that which is independent and dependent God is wholly independent and the world is entirely and always dependent on God. God is infinitely perfect.

3. *Matter and Spirit.* The world is made up of two kinds of substance, matter and spirit, material things and souls. The individuality and uniqueness of each and every thing is an obvious fact of experience

We can understand the relevance of Madhvacharya's insights here by pointing out where he differed from Buddhists and Hindu monists. Unlike these two thought-forms, he affirmed that:

♦ We really do exist

♦ We have a consciousness and an individual identity that we will retain permanently

♦ We can know things

♦ God exists and we are distinct from and dependent on God; God has attributes that can be known

♦ The ultimate goal of life is union with God, a union in which we retain our distinctive identities.

In studying Madhvacharya's insights and especially his comments on monism, our primary source document is B.N.K. Sharma's *A History of the Dvaita School of Vedanta and Its Literature.*

How We Know

At the basis of science, philosophy and theology is the question of whether and how we can know anything. The easy way out here is to play the skeptic and deny that we know anything. But this would not be true to experience since we do know and know we know.

Madhvacharya gave an original answer to this question based on what is obvious in daily life. He held that it was possible to secure valid knowledge ("yathartham") through one of three channels:

+ Accurate perception with the senses

+ Flawless reasoning (valid inference)

+ Divine revelation (verbal testimony).

Through any one of these three means, one arrives at knowledge that corresponds to what is real, to truth itself.

Although he holds that truth not falsity is the norm in experience, Madhvacharya admits that it is possible to be in error in judgments based on sense-experience, reasoning or the interpretation of revelation. It is here that he makes an original contribution by drawing attention to a deeper dimension of our being, the experiencing self or *Sakshi*. Madhvacharya believed that in addition to knowledge that comes through the senses and reasoning, our inmost self has a fundamental ability to discern truth from error that he called *Sakshi*. This ability is required for us to know anything to be the case. Thus, "the very validation of knowledge depends upon the *Sakshi* which is the ultimate principle that knows the knowledge." It is *Sakshi* that is the primary guarantor of truth and certainty in our coming to know something. *Sakshi* is a truth-determining principle that discerns truth from falsehood, and without *Sakshi* one cannot know anything to be the case. When applied, its judgments are self-guaranteeing or self-luminous.

"The mere awareness of [the] knowledge [of something] does not reveal its validity," wrote Madhvacharya "Such validity is not realized at the very outset, in sensory and other forms of knowledge. It is only when it is intuited by the *Sakshi*, with or without the aid of tests, that the true nature of validity, comes to be clearly and fully realized and manifested. It is therefore absolutely necessary to draw a distinction between ordinary sensory, mental and other...[states of mind] and the judgments of *Sakshi*, in accordance with facts of our own experience."

God and the World

It was obvious to Madhvacharya that there is a radical difference between God, the only truly independent being, and all other things which are dependent on God and preserved in existence by him. God is perfect in being and infinite in the sense of not having limitations of any kind. He has no dependency whatsoever and is entirely self-sufficient and free of all imperfections. Madhvacharya writes, "Whatever would be incompatible with the sovereignty of God should be rejected. Inconsistency with divine majesty would itself be the criterion of what is unworthy of acceptance. All proofs and authorities should be interpreted in the light of this criterion, that the supremacy of God should not be compromised."

Madhvacharya argues that two equal independent ultimate principles, two Gods, would be impossible. There can only be one independent reality with all else subordinate to him. When God is called the One without a second, we are referring to the fact that he exists in his own right with no dependence on anything else.

In his commentary, Sharma draws attention to Madhvacharya's affirmation that God is not featureless as the monist *advaitins* say, but has attributes that can be known and experienced. Brahman "should possess attributes like omniscience insofar as it is the creator of everything in the universe," affirms Madhvacharya. "The all-creator must be all-knowing, all-powerful and capable of accomplishing whatever He wills."

Thus, God is omniscient, knowing all things; omnipotent, all-powerful; eternal; transcendent; immanent; and omnipresent, present everywhere. God is the source of all else that is real, conscious or active. God is creator and preserver of the world; souls are dependent on him and are delivered from bondage by his grace. There is no distinction between God and his attributes, activity and will. In fact there is an absolute identity between God and his attributes and between each attribute. God is personal because he is self-conscious and capable of cognition and action. Madhva also showed how the personhood of God relates to his moral and metaphysical attributes.

Although Madhvacharya never considered the idea of *creatio ex nihilo*, he maintained that no finite thing could exist without God. The existence, knowability and activity of the world are dependent on God. God is the efficient cause, the maker or producer of the world. The world and all the souls are metaphysically dependent on God, who is the only independent cause. God controls all the activities of nature and energizes souls. God is the Original and the world is a Reflection. Madhvacharya denied that the world is either a part or a transformation of God, and avoided all forms of pantheism. He was equally adamant in rejecting the idea that the world is an illusion.

Matter and Spirit

Madhvacharya recognized that there is a qualitative difference between matter and the human soul (*Jiva*). Every *Jiva* or soul is an indivisible unique center of consciousness, a self that can experience and know, a doer who chooses freely. No soul can share the immediate experience of another. Moreover, the contents of the consciousness of each soul are different from every other, as is its state of bliss, knowledge and virtue. In these and other respects, every soul is unique and will always remain distinct from every other soul and God. The goal of life is eternal union with God through self-surrender, but Madhvacharya emphasizes that such salvation would be worthless if the self does not retain its individuality in this union. Union with God brings bliss, but there would be no bliss if there is no one who experiences and enjoys it.

Critique of Monism

Madhvacharya's critique of monism, particularly as embodied by Sankara and his philosophy of *Advaita* Vedanta, centered on four areas:

1. Our experience of differences and of the world as real

2. The incoherence of pure consciousness

3. The incoherence of saying that God has no attributes

4. The misinterpretation of the Vedas and Upanishads

We will review Madhvacharya's critique in our study of monisms of the East.

Chapter

An Eight-fold Path to a Theory of Everything

*T*HE STARTING-POINT OF OUR INQUIRY was the integral connection between the Matrix and the foundations of science. We have seen that there
are at least some themes in common between the two. I have argued
further that only the Matrix as a whole can explain the assumptions and
findings of science.

So how is the Matrix relevant to science today? It seems to me that the
Matrix is a meta-scientific Theory of Everything (TOE). This TOE is made
up of pre-scientific and even pre-philosophical truths that are assumed
by both science and philosophy. Although scientists and philosophers can
discuss these truths, they cannot prove or disprove them with their tools
of trade. These are truths that have to be seen, perceptions of the inmost
self. Here we are dealing not simply with thoughts too deep for tears but
insights too deep for thoughts. The TOE is important because it enables us
to go beyond the data of science to its ultimate meaning.

A meta-scientific Theory of Everything can also be called a world-view
because it is an intellectual framework that makes sense of our experience
of the world. The Matrix as a TOE meets the three criteria required to establish its credibility as a world-view:

1. *Foundations that are undeniably true.* I hope to show that these foundations are indeed self-evident truths.

2. *Logical coherence*, unlike spiritualist monism. The monist says
    there is only one reality that exists; but, if this assertion is true,
    then there are two realities in existence, both the Ultimate Real-
    ity and yourself, since there is obviously a "you" who thinks the
    thought and makes the claim.

3. *Comprehensiveness of explanatory power.* The Matrix accepts the
    physical, mental, moral and religious dimensions of our experience
    on their own terms; materialist monism denies or explains away
    the last three.

By the very nature of the case, a TOE of this kind is not affected by changing scientific hypotheses and theories, since it's exclusively concerned with
the structure of reality presupposed by science.

My present task is to lay out eight foundational principles that are an embodiment of the Matrix for our time. It is, if you will, an Eight-Fold Path
of thought that leads us to a modern version of the Matrix, the meta-scientific Theory of Everything. The principles laid out as the Eight-Fold Path
are implicit or explicit in the Matrix formulated by the Four.

What do I mean by a meta-scientific TOE? In current discussions, a Theory
of Everything is a theory that comprehensively describes everything that
exists in terms of one or a few simple laws. The theory would not only
explain how the universe came into being but why every part of it, from
particles to galaxies, have the properties they do. It unifies all the laws of
nature, all fundamental constants and interactions, under one umbrella.
Some believe that M–Theory could turn out to be the true TOE.

TOES are concerned with the physical world and are therefore scientific. But
a scientific TOE is extremely unlikely because of Gödel's famous Theorem.
The Theorem states the obvious fact that every logical system has certain
truths that cannot be proven within it. The consistency of axioms assumed
by the system cannot be proven within it, but have to be proved within
some more fundamental logical system that will have its own axioms that
in turn are not provable within it, and so on without limit. Physicists like
Paul Davies, John Barrow, and, most recently, Stephen Hawking, believe
this theorem dooms all hopes of a TOE. As Russell Stannard has pointed
out, a TOE would have to show that only the particular universe in which
we live with its specific set of laws would be possible, but then we would
have to generate a mathematical model of the universe, and per Gödel this

model would have axioms, and their corresponding physical laws, that can't be explained within it.

Another critic of scientific TOEs, Mitchell Feigenbaum, a pioneer of chaos and complexity theory, says incisively that many of his colleagues "like the idea of final theories because they're religious. And they use it as a replacement for God, which they don't believe in. But they just created a substitute." Which brings us to the idea of a meta-scientific TOE.

A theory of everything is meaningful only if we really take everything into account. And everything includes not just fields and universes but rationality, life and mind. And, rightly understood, these categories involve realities that go beyond purely scientific tools. A scientific TOE needs a meta-scientific counterpart. Whether or not a scientific TOE can ever be found, it's certainly possible to sketch a meta-scientific version that addresses the question of Ultimate Reality. But what tools do you employ in the latter? How do you arrive at meta-scientific truths?

Clearly, we have to sharply distinguish between scientific and meta-scientific/ontological questions and claims. Much of the confusion surrounding discussions in these domains arises from simple conceptual muddles. Science deals only with things that are measurable, quantifiable and repeatable. Quantities and the measurement of quantities are what science is all about. Ontology, which is the domain of meta-scientific questions, is concerned with principles, insights and truths that by their nature transcend scientific theory and practice. These include self-evident truths presupposed by science and issues that cannot be addressed with any scientific methodology. *If we ask what are the laws that govern the universe, we are asking a scientific question. If we ask why does a structure of laws exist, we are asking an ontological question.* The data of science can, of course, serve as the starting-point for ontological study but that study will require ontological and not scientific tools.

Now certain scientists might respond that they're only interested in cold hard facts, not so-called meta-scientific or ontological ones. But it's easy to show that even the most hardheaded experimentalist can't get away from the ontological realm even for an instant. I ask: How do you determine that something is a "cold hard fact?" You make a mental estimate by weighing the evidence for and against, and you try to find out if the premises warrant the conclusion or if known facts support the hypothesis.

All of these mental acts are ontological judgments. You can't arrive at a judgment by pouring the facts into a test-tube or peering at them through an electron microscope. So even to do "hard" science, to generate, evaluate and categorize data, you need to go beyond hard facts and concrete reality. What you need is a framework of "cold hard truths," and that's what you get in the Matrix and its proposed embodiment here.

Of course, meta-scientific issues fall outside philosophy as well. Basic ontological truths cannot be derived from elaborate philosophical or logical argument. The conclusion of a philosophical argument is already contained in its premises. If you accept the premises, the conclusions follow. But why do you accept the premises? Ultimately, these premises have to rest in self-evident truths or hard facts of experience. As I will try to show in my discussion of how we know, all of us know the most important truths already, although these may be hiding in the backs of our minds obscured by confused thought-forms.

The task at hand is to retrieve this great treasure house of truth buried in our minds. Many pop philosophies and esoteric thought-trends of our time are built on ideas that can easily be shown to be errors. If the basic idea is shown to be wrong or, even worse, incoherent, then the superstructure built over it collapses. At the same time, we can know some things to be true. Many of our contemporaries think that there's no way of knowing what's the case about anything because there is disagreement on everything. This is an age-old temptation. If you ask ontological questions using scientific methods and concepts, of course, you will have a cacophony. But once you put a finger on the turn wrongly taken, you can commence your journey with better directions in hand. And it can be shown that there are some hard facts that underlie both science and philosophy that can be denied only at the risk of self-contradiction. These are the hard facts that comprise the meta-scientific Theory of Everything.

In searching for these particular hard facts, our objective is to discern and describe those principles that underlie what is most obvious to us. So what's obvious?

- There's the existence of laws in the natural world that (a) are very precise, (b) give rise to structures, and (c) enable life.

- There is the intrinsic purposiveness of all the systems and processes we see around us (eyes are for seeing, reproductive systems are for reproduction, gluons carry the strong nuclear force). There is a qualitative difference in intelligence between living and non-living systems.

♦ Then we have the human person and the whole world of being conscious, having thoughts, retaining self-identity despite continuing physical change, making choices.

What we need is a paradigm, a framework of thought, that plausibly and coherently accounts for and accommodates all these hard facts.

So here's my line in the sand, the essential and ultimate principles that make sense of the world as we experience it. The rationale for these principles will be presented in our dialogues.

1. *First* and most fundamental, Garbage In=Garbage Out: your output is equivalent to your input.

2. *Second*, What You See is What You Get: to be acceptable, any explanation of the past must tie in with what we experience in the present.

3. *Third*, something cannot come to be from pure nothingness.

4. *Fourth*, an intelligent system cannot be a product of inert matter.

5. *Fifth*, life in its essence is so different from non-life that we cannot conceive of a non-living system spontaneously generating life.

6. *Sixth*, the capabilities of the mind, consciousness and conceptual thought, are so radically different from the physical world that we cannot coherently think of matter producing the mind.

7. *Seventh*, the self understood as a center of our consciousness, the unitary unifier of our experiences, is radically different from things that cannot be thought of as having a self, although category confusion leads us to talk of self-organizing matter or the selfish gene.

8. *Eighth*, the intelligence we observe at all levels of the universe, inanimate and animate, human and non-human, leads us to recognize that there is an infinite Intelligence that is the source of all the intelligence in the universe. Concurrently we recognize that infinite Intelligence is not and cannot be sub-personal since it's simply incoherent to think that the personality and selfhood we experience for ourselves can come to be from a mere force field.

None of these eight principles can be proved or disproved by science or scientific methods since in some cases they are presupposed by science and in others deal with realities not subject to physical observation or measurement. They are ontological principles that we come to recognize as self-evident truths once we reflect on what they mean. We see each one of these principles to be the case. Seeing is the terminus of the quest for truth. Knowledge, said Illtyd Trethowan, is basically a matter of seeing things, and reasoning processes can get under way only if we have a starting point in direct awareness and apprehension.

The truth of these principles becomes evident simply by reviewing attempts to explain all things in terms of matter. Just when the drumbeaters for science thought that they had eliminated all mystery, modern cosmology revealed a universe that was expanding and inflating, awash with quantum energy. Just when initial research on the origin of life seemed to indicate that scientists were close to creating the building blocks of life in the lab, it became evident that the DNA and the proteins fundamental to life were too complex to emerge from random reactions; there was no choice but to postpone the problem by appealing to panspermia (life came from outer space) or "just so" concepts like complexity theory (life simply emerges from unpredictable interactions). Just when the materialists thought that they had constructed a fairly convincing account of the mind as a material entity, a whole new generation of scientists admitted that consciousness was a mystery and a challenge that had to be taken seriously by science.

It's amazing how the frame of mind of the materialist monist parallels that of spiritualist monists. They know everything there is to know. Anything that doesn't fit in is ignored or explained away. But this arrogance shouldn't impress or intimidate the sincere inquirer. We will press the questions home regardless, knowing that we can't deny the obvious or suppress facts. Can the scientist who seeks to explain the workings of the universe explain the workings of his or her mind, the relentless passion to know, or the ability to form concepts and make deductions?

Remember, I have no desire to downplay the staggering output of scientific data. In fact, it's science that bears the strongest and clearest witness to the rationality and wonder of the world. The more data it uncovers, the deeper the mystery and the greater the awe.

Let me start with what seems to be obvious. Garbage In, Garbage Out (GIGO) and What You See is What You Get (WYSIWYG) are two principles that are accepted by most computer programmers. I believe these two principles have a wider application than you might initially imagine.

1. Garbage In = Garbage Out

Take GIGO. Among the computer literati, this generally means that a system can only produce what a programmer or data entry technician has instructed it to produce. Output=Input. In the real world this means that no property can come from a being that does not already have this property or the power to bring about such a property, i.e., *A* cannot on its own produce *B* if *B* was not somehow contained in *A*. All the matter that makes up a baby calf comes from its parents, and environmental influences come later. All the elements that comprise a work of art or music come from its author and the medium used by that author. The capabilities of computers and other mechanical devices, no matter how sophisticated, were implanted in their entirety by those who made them out of pre-existing raw materials. Every one of these end-products—calf, artwork, and computer—was present as a potentiality in their originators. Conversely, this principle stipulates that you can't produce something that is not somehow already present in you; likewise, a collection of things or systems can only produce what is collectively present in them. Rocks can produce pebbles, but not flowers or minds.

Does this mean that genuinely new things cannot come into existence? Well, it seems obvious that all that exists is in some way a product of some other thing or combination of things that existed previously. Even the most determined materialist would base his or her ideas of origin on some theory of how matter gradually evolved into life forms and the diversity around us, while also holding that these can all ultimately be reduced to matter.

Now what are the implications of GIGO for our current discussion? A few prominent examples may be considered. Life cannot originate in a non-living object unless that object already in some fashion contained life. You might say that certain non-living objects interacted with the environment to generate life, in which case you assume that between them these objects and their environment contained *all* that constitutes life. Either life came from outside or it was present inside its non-living precursors. Or take the intelligence in the world, the intelligence we experience in ourselves, and that is manifested in the natural world. Could intelligence originate from mere matter? Is it reducible entirely to matter? No matter how sophisticated or refined, a theory of materialism tells us that the world with all its wonders is purely and simply the product of a vacuum fluctuation or some such physical reality. The order and symmetry in the universe as well as roses and Shakespeare and Mozart were all present in inert matter at all times. When applied to neo-Darwinist theories of evolution, GIGO tells us that you can mutate all you want but you can't create what's not already there; what you mutate into must already be present. I for one find it impossible to believe that life, consciousness and mind were always present in matter.

2. What You See Is What You Get

My other metaphor, WYSIWYG, the ability to develop a software application or web page without having to do any coding, is analogous to a fundamental principle in Darwinist theory: "The present is the key to the past." I heartily agree with this principle but would insist that we apply it across the board. In brief, all speculation should be firmly anchored in what our present experience tells us. Any explanation of the past must adequately explain what we experience in the present: the intelligence embedded in all things, the pattern, structure and law in every object we observe; the vivid and unmistakable experience of our personal identity, the self; and our equally undeniable awareness of being conscious.

3. Something from Nothing

Thirdly, absolute nothingness cannot produce something. By nothing I mean precisely that: no laws, no vacuums, no fields, no physical or mental entities of any kind. It is simply inconceivable that something could spontaneously pop into existence out of sheer nothingness. We can force ourselves to believe that this has happened if we want, but we will have no evidence to support the belief. Moreover, once we understand what nothing means, it makes no sense to talk of something coming to be out of nothing.

This third of the eight principles is perhaps the one that is most obviously true, but curiously, the modern mind seems oblivious to what is obvious. Thus, we often hear that quantum physics or quantum cosmology has shown that—yes—nothing has indeed produced and continues to produce something. But this quantum appeal in support of the idea that nothing can produce something is based on linguistic or conceptual confusion. Over the centuries thinkers who have considered the concept of nothing have been careful to emphasize the point that nothing is not a kind of something. Absolute nothingness can never be the object of scientific inquiry because all such inquiry presupposes the existence of the object of study and of some order governing the behavior of the object. Science can only get to work once something exists. Consequently, the transition from nothingness to something lies forever beyond the purview of scientific methodology. The nothing that contemporary cosmologists and quantum physicists discuss always turns out to be something in disguise.

If we grant the truth of this principle, we might then ask how it is possible that anything at all exists. Why is there something and not just nothingness, and where did the something come from? At this point we are asking a meta-scientific and not a scientific question. Science can tell us about the things that are already in being around us, the laws that govern them and their various transmutations. But science cannot tell us how the material reality it studies came into existence, just as a study of our ability to form various ideas and concepts cannot tell us how we came to have the ability itself.

A cosmological theory like the Big Bang is a brilliant attempt to track the evolution of the material universe back to its earliest moments, but this program is irreducibly different from the attempt to answer the question of how it is that anything at all exists. In our study of modern cosmology, we will address various scientifically fruitful hypotheses about the origin of the universe. But the origin of the universe is not the same thing as the origin of being. Once you concern yourself with the ground of material reality, you have moved beyond science into ontology.

છ૭

4. Intelligence from Non-Intelligence

The fourth principle is that inert matter cannot produce an intelligent system. This is in some respects the most controversial of the eight principles, because most people assume that Darwin has put it to bed with the ideas of random variation and natural selection, and, if not Darwin, some of the self-organizing scenarios that dominate contemporary discussion. But, as a matter of fact, this principle is compatible with both natural selection and the various self-organizing models. Both views assume that something exists and that this something follows certain laws of nature. The question is really about the system of laws that predate natural selection.

Alternatively, you might argue that the universe popped out of a quantum vacuum or that matter always existed. Even in such scenarios, you are left with a vacuum that follows certain laws or matter that has certain properties, and it is the origin of these laws and properties with which we are concerned.

I must confess I cannot see how anyone can seriously propose that all the intelligent systems we see in operation here and now are ultimately random products of a vacuum fluctuation or of eternally existing matter. How is it possible for a system of precise organically unified laws and constants to spring from absolute chaos or inertness? Scientists have proposed elaborate probability models to explain why under some scenarios, improbable but nevertheless possible, the conditions for life would be present in at least one universe. But that is not the point at issue here. Our question concerns the origin of the very laws that would make any universe possible. A lump of clay cannot give birth to a working computer given an infinite period of time or an infinite number of universes. This analogy is imperfect, but only because certain laws that are inexplicable in themselves govern even the clay.

One objection that has been raised is self-organized criticality. Grains of sand randomly dropped on a pile follow what are called "power-laws" without any external intervention. Again, it is said that computers that simulate the role of chance have given plausible reconstructions of the random origin of life. But does the sand follow certain internal and external laws? Do the computer models assume that certain hardware and software configurations are in place? Of course they do. We are asking how these laws came to be. Has anyone devised an experiment where a law of nature springs into existence from anarchy? A transition from chaos to intelligence is in principle impossible.

5. Life from Non-Life

Regarding the fifth principle, I admit that most chemists and biologists think that the origin of life is simply a biochemical question. Of course, life as we know it has a biochemical dimension. But can life or its origin be reduced solely and simply to biology and chemistry? I think not. In my discussion on the origin of life, I will try to show that (a) all living beings, from microbes to plants to animals, are autonomous agents; (b) this agency involves self-motion, self-generation, self-replication, qualities that cannot be reduced entirely to unintelligent matter; (c) all life centers on DNA, a complex reality that arose fully formed several billion years ago and has stayed the same ever since; and (d) there is a radical distinction between life such as we find in unicellular organisms and the conscious life of animals. Although attempts have been made to blur the distinction, it is a fact that we have no conclusive or universally accepted scientific evidence or mechanism for the transition from the basic forms of life to the conscious sensory life of animals and finally the rational life of human beings. The question we will address is this: how did life with its intricate hierarchy originate?

6. Mind from non-Mind

What is the mind? I would say the human mind is the self-conscious center of rational thought. Just as materialists reduce life to a biochemical accident, so they say the mind is solely and simply the human brain, which is in turn a product of evolution. I will try to show that the materialist position is demonstrably false.

7. Self from Non-Self

Closely related to the notion of the mind is the idea of the self. Whatever else you wish to dispute, you can hardly deny that YOU, the person I'm addressing, have existed continuously, through numerous cellular transformations and changes of brain state, as the same individual from the date of your birth to today.

As you know, I can't accept the coming to be of a mind from matter. Even less so am I willing to accept the incoherent idea that a self could emerge from any kind of primordial stuff. If we are centers of consciousness and thought who are able to know and love and intend and execute, I can't see how such centers could come to be from something that's itself incapable of all these activities. It's simply inconceivable that any material process or field can generate agents who think and act.

Now it might be said that there's no "I" separate from my body and nervous system. But this approach is totally untenable if you want to do serious science or just live in the world. Your every intention and statement presupposes the existence of someone who's doing the intending and the stating. Now you might say this "I" is a conglomerate of cells, but then I ask what's the center, the hub of the conglomerate. You say there's no such center or that certain brain cells function in this role. There are a number of mini-agents, your cells, but no overall super-agent that runs the show. At this stage, I remind *you* that *you* constantly experience the clear, indubitable awareness of being an "I," of being not just conscious but self-conscious. Of course you could say that your sense of "I," your self-consciousness, is just an illusion. But an illusion has to be experienced *by someone* to be an illusion, which brings us back to the existence of a center that is conscious and rational. You, the person, cannot be found inside your brain or nervous system for the same reason that you can't attribute a murder purely to the murder weapon, which is simply the vehicle used to actualize the intention of a person. The self can't be reduced to chemicals or be produced by them.

I find it intriguing that those who vociferously deny its existence somehow end up importing the idea of a self into the most unlikely places. We hear of the "self-organizing" properties of matter, or the ambitions and actions of the "selfish gene." Of course, it should be obvious that there's no self that's capable of cognition and consciousness involved here, and thus the application of the term "self" in these contexts, amusing as it may be, is misplaced and misleading.

෧৯

8. Infinite Intelligence

The principles outlined here raise the question of origin. If something cannot come from nothingness, life from non-life, mind or self from a material matrix, where did the world and its life forms come from? My thesis in the beginning had been that embedded, active and self-aware intelligence could only be explained by reference to the existence and immediate activity of infinite Intelligence. In a separate presentation, I will address the question of the existence and nature of God. Here I will simply say that mind can only come from infinite Mind, consciousness from an eternal Consciousness and the self from the supreme Self. Given a trillion years or forever, fields or particles cannot generate thought. Nor can absolute nothingness ever become something.

We call this Intelligence infinite because it is Intelligence without any limitation: it knows all things, knows all there is to know. To be infinite is to be free of any limits. How do we know that an infinite entity in this sense exists? Simply from the fact that anything at all exists. The recognition that our existence is unintelligible if it is not grounded in a self-existent God is also the recognition, without knowing *how* this is possible, that God exists always with an existence that is free of all limitations. Any being that exists by its very nature has all the perfections of existence. This is what it means to say that God is infinite and perfect. God is the perfect fullness of reality that not only possesses all transcendent perfections but also possesses them in an unlimited degree. The divine infinity, inasmuch as it signifies a freedom from any limitation, entails that God is not limited by either space or time. But these are all areas we will consider down the road.

In walking down the Eight-Fold Path laid out here, we realize that we're exploring areas in which we rely only on what we can see for ourselves. Intellectuals are as prone to fashion and fad as rock music fans and clothing designers. Establishment dogmas and orthodoxies are subject to change without notice or rationale. They never last beyond a generation, if that. All of us must die, and there is no unchanging collective memory that preserves a scientific or philosophical ideology for much longer than the lifetime of its creators. We can try to pander to the intellectually respectable or politically correct fads of our time, but these are passing fashions that have very little to do with the truth about things. Life and death, origin and destiny are deadly serious matters, and we owe it to ourselves to ponder them with a critical eye and an open mind. We are on our own when it comes to making judgments on the facts of life.

Concretely speaking, this means that we can only accept affirmations and claims that are plausible on the face of it, and that the burden of proof will quite clearly have to rest with those who take counter-positions. If someone says that time doesn't exist and our experience of change is an illusion, the burden of proof is on them to prove their point. Philosophers and scientists have seriously proposed this view.

Given the stakes, we have no choice but to deal only with hard facts and self-evident truths. How we discern what is self-evident is a topic I will address as we proceed. Speculation, guesswork, probability calculations all have a place in certain areas, but in considering the world as a whole and humanity in particular, we will only take seriously what is undeniable or irrefutable. This is what the Matrix is all about. And this is all that I'm proposing in the Eight-Fold Path.

THE

2

GURU GEEK

DIALOGUES

PART I
Seeing Is Believing

Chapter

Triad

GURU: Well, I'm sure you thought my monologues would never end. But now I'm ready for the next phase of our journey. You've seen my vision of the world and my view of how science came to be. From our discussions in the last few weeks, it seems to me you've come a long way from the standpoint you adopted at the conference. While you still believe that the world we live in is ultimately nothing more or less than matter in motion, you're willing to concede that alternate views are worth considering.

The materialist monist says that all that we see and love, all that we are and have, are the accidental products of physical fields at one level and of random mutations at another. My position is that, *au contraire*, modern science has introduced us to a world of wonder that manifests the presence and action of infinite Intelligence at every level of its being. More important, I affirm that we encounter this infinite Intelligence in everyday experience. To put it plainly, as I've said, we see God at every waking moment! Of course, when I talk of "seeing," I'm not talking of the direct and immediate vision of God that some traditions call the Beatific Vision and others *moksha*. It is more a case of seeing a dim reflection in a mirror. But it is seeing nonetheless. Granted, you can't see God if you can't see at all, and that is a bridge that has to be crossed first.

Geek: I'm ready and willing to listen. So show me!

Guru: Wonder is the beginning of wisdom in this case. The proper pathway to a perception of infinite Intelligence is the wonder that is the world. This is the magic and mystery of which the poets, philosophers and scientists have written. But our quest will not take us to esoteric enclaves or mystical monasteries. We are, instead, journeying to the heart of everyday experience. For the most obvious things are the ones most easily missed, and the truth about things cannot be any further from us than the air we breathe or the ground we tread.

> *Geek*: I have to confess that I have yet to find anything wonder-filled in my everyday experience. I'm not opposed to searching and finding the wonder-filled. But it's certainly not obviously present in my case.

Guru: Thanks for being honest. This is, in fact, the condition of mind I run into almost everywhere. It might be called the collective coma of the modern consciousness. Please don't take offence at my description since it's simply a general observation. I hope that our dialogue will help me show you the obviousness of what you currently consider opaque.

> *Geek*: In the spirit of coming right to the point, let me say that what's sauce for the goose is sauce for the gander. Those who're diagnosed as suffering from a "collective coma" can justifiably ask why they should believe that you're somehow mystically in touch with things in a way that mere mortals aren't.

Guru: A fair question, to which I hope to give a satisfactory response. My whole approach here will be one that conforms to the ancient dictum "seeing is believing." I think this is a starting-point on which we can both agree. I hope to avoid pedantic pontifications. I hope to give you as much evidence as there is available, and it's more than sufficient in my view.

> *Geek*: Well I guess it all depends on what we're going to see and what kind of belief is said to follow from the seeing.

Guru: Our journey begins and ends with the Triad of Creativity, Rationality and Energy that constitutes the world. To make any progress we have to act as if we are seeing everything for the first time.

First, we have the sheer *inventedness* of all things: everything from a blade of grass to a galaxy is a unique reality that somehow came to be here. The first cookie and the first automobile were ideas in the minds of persons who then actualized them. Where did the ideas of electrons and neutrinos and cells and genes come from, and how were they turned into things?

Second, we encounter the irreducible *rationality* underlying the physical world. At every level, this world is a realm of complex structures governed by precise laws. This world-wide web of information-rich process-flows, just-right physical constants and all-encompassing symmetries is something I call embedded intelligence.

Third, we see the entire cosmos bursting with *energy*, throbbing with activity. From subatomic particles to the Big Bang, from photosynthesis to the supply chain of the cell, the momentum of being is a constant fact of experience. What is the source of this dynamism that powers all things and what keeps it going?

The Triad manifests itself most fully in this world through two key agencies. The first is *life*, the ineffable but indomitable attribute that differentiates a stone from the moss that grows on it, a starfish from a star, a lightning bug from lightning. From metabolism and replication, the fundamental characteristics of all living beings, to locomotion and consciousness, the capabilities of the higher animals, life inaugurates a new horizon in the universe. The second is *Homo sapiens*, thinker and actor, planner and executor. From self-consciousness to freedom of the will, from language to love, *homo sapiens* embodies a mystery more profound than the entire universe.

> *Geek*: Wouldn't you agree that these phenomena have been catalogued and exhaustively analyzed and explained by modern science? To be sure, there are some anomalies and puzzles, but given the exponential pace at which yesterday's mystery becomes today's triviality, it seems short-sighted to appeal to a god of the gaps to sort out the few remaining quirks in our picture of the world.

Questions that seemed profound in the worlds of the ancient Greeks, Indians and Chinese and the medieval Europeans are irrelevant, even meaningless, in the world of Darwin and Einstein and quantum physics. For example, the quantum cosmologies of today seem to have eliminated any remaining questions about the origin of the world. Again, an emerging Grand Unified Theory could well explain why everything is and has to be the way it is.

Guru: First, I want to distance myself from any attempt to bring in a god to explain the so-called gaps in strictly scientific descriptions. We will come to the question of God later, but let me say here that we won't find him in a loophole. In point of fact, it's mostly modern physicists who talk about God in the context of new discoveries, although their understanding of the concept is sometimes anthropomorphic.

Second, far from closing them, modern science has only widened and deepened the gaps that are ontological and not scientific. The more we learn about the world in modern genetics, particle physics and cosmology, the greater the mystery and, yes, the *wonder*.

Third, the idea that science will somehow eventually answer all questions is a reverse version of the god of the gaps argument. Rather than seeing science as a battering ram that bludgeons all domains of knowledge into submission, I see it as a marvelously effective methodology for broadening and deepening our understanding of one dimension of reality, the physical. But science itself rests on facts that lie forever beyond its reach. For instance, by its very nature, science can tell us nothing about such questions as the origin of the physical or the nature of conceptual thought. Of course, some self-proclaimed evangelists of science have broadcast theories that purport to scientifically explain every domain of human experience. But the very fact that these theories fall apart in rapid succession is an illustration of their inherent absurdity. Their authors are simply tilting at non-quantitative windmills with tools built for measuring physical quantities. The graveyard of the gods of the gaps was populated in the past by superstitions rooted in religion; almost all the recent tombs, however, belong to deities with distinguished scientific credentials. For instance, Pasteur experimentally refuted one version of the idea that life could simply spring from matter given enough time. Again, Thomas Huxley's claim to have observed an organism that was halfway between living and non-living was accepted by scientists; on further study, it turned out that this "organism" was simply a mineral.

The same questions that puzzled the Greeks, Indians, et al., are still with us today. Darwin, Einstein and many prominent quantum theorists admitted the intractable nature of these questions, and the modern world has come no closer to answering ultimate questions than the ancients. With regard to GUTs and theories of everything, the biggest obstacle is Gödel's Theorem, an area we will presently cover.

Geek: Let's move on to something specific and tangible.

Guru: To reiterate the point made earlier, I'll proceed on the principle that "seeing is believing." An affirmation is acceptable only if the evidence for it is conclusive. And, we can't decide what is to be believed until we agree on what it means to see.

Geek: I'm intrigued.

ॐ

Chapter

2

Seeing with the Brain

*G*URU: The skeptic likes to say, "I have to see it to believe it." But what is seeing? To understand the act of seeing is to take the first step towards the wonder of the world. It is the datum that launches a thousand trains of thought. We will consider seeing at two levels, (1) in terms of the physical act itself and (2) in the sense of grasping the meaning of something or knowing something to be the case.

You might ask why I'm starting our dialogue with the topic of seeing. Well, there's a very good reason. The last of the Seven Wonders was about seeing God. As I see it, this is the destination of our journey, but before we can see God we have to know what seeing itself is all about. In a sense, seeing is the key to everything else. Its very naturalness might tempt us to take it for granted. But, at a physical and mental level, seeing is perhaps the most mysterious phenomenon in our experience. And, once we've seen seeing for what it is, we will be better prepared to address the intelligence and rationality in nature as a whole.

Physiologically, the process of seeing involves:

1. The existence of light with which we see.

2. The existence of a highly complex organ, the eye, capable

of processing visual information including color, shape, distance of external things.

3. The actual physiological processing of visual information through various structures in this organ.

4. The transformation of visual data from electromagnetic vibrations to nerve signals in the brain to perceptions in the mind that are subjective sensations of shape and color and motion.

Almost everything about this process is baffling:

♦ How did this intelligent organ with its complex capabilities come into being?

♦ Why is there a perfect "fit" between the structures of light and the eye?

♦ How do we explain the peculiar relation between eye and brain and mind?

Before you dismiss these questions as pre-scientific paradoxes, let's take a long, hard look at the physiological facts and then the questions they raise.

The human eye is made up of a number of structures. For our purposes, the most significant is the eyeball with its three coats of tissue:

♦ The outer protective tissue is the sclera, and its transparent center is the five-layered cornea that refracts light.

♦ The choroid is the middle layer of vascular tissue connected to the ciliary muscles and processes in the back that provide various nutrients to the eye. In the front, the choroid changes to pigmented muscles called the iris, at the center of which is the opening called the pupil.

♦ The innermost coat, the retina, is a nerve tissue that is connected to the optic nerve that in turn leads to the brain. At the back of the retina is the transparent, light-sensitive coat that receives images of the world.

Behind the iris is the lens, a transparent body of fibers enclosed in a capsule. A watery fluid of blood plasma called the aqueous humor provides oxygen and nutrients to the lens and the cornea, and fills the space between them. Behind the lens, enclosed in a sac, is a jelly-like substance called the vitreous humor.

The act of seeing involves the brain as much as the eye. The eye converts the electromagnetic vibrations that constitute light into nerve impulses and then sends them to the brain. Much like a camera, the lens of the eye creates inverted, smaller-than-life-size images of exterior objects on the film in the back we call the retina. When we look at an object, light reflected from it passes through the cornea, the aqueous humor, the pupil, the lens and the vitreous humor to finally reach the retina. The iris adjusts the size of the pupil depending on the amount of light coming in. The lens thins or thickens as necessary to allow each ray of light to focus on the retina. Then the rays of light are bent so that they converge at the center of the retina. The impulse that reaches the retina is then transmitted to the million-fiber optic nerve and from there to the occipital lobe of the brain where electrical signals are seen as right-side-up and life-size visual images.

The retina, which is about one centimeter square and one-half millimeter thick, has 100 million neurons that are functionally divided into five classes: photoreceptor, bipolar, ganglion, horizontal and amacrine cells. As it carries out its image analysis, retinal neurons communicate with each other through synapses using chemical messengers called neurotransmitters. The photoreceptors, made up of millions of rod and cone cells, are used by the retina to distinguish color; the rods give the intensity of light distinguishing between light and dark while the cones read its wavelengths, red, blue and green in particular. The color of a thing depends on which of the various colors that make up the visible spectrum it reflects or absorbs or transmits. A chain of chemical reactions is triggered off when a retinal cell absorbs light. A single photon acting on the rhodopsin protein in a rod cell sets off cascades of enzymatic activities that translate this information into a signal that can be processed by the nervous system. The retina processes ten one-million-point images every second. To match the retina's processing power, a robot vision program would have to perform 1,000 million computer instructions per second. It has also been estimated that computer simulation of the processing performed by just one retinal nerve cell in one hundredth of a second would call for the solution of 500 simultaneous non-linear differential equations 100 times.

❧

Geek: Thanks for the optical tour that I hope won't be an optical illusion. If I'm not mistaken, the whole point you're trying to make is that profound mystery surrounds the origin of the eye. But isn't this just one more temporary "gap" that doesn't need a *deus ex machina*? Let me caution you that this alleged mystery is no longer seen as such by hundreds of scientists. The biologists say there is a clear progression in the capacity for sight in the history of evolution, beginning with organisms with light-sensitive cells that can detect the presence of light and darkness. Through millennia, driven by adaptation, these cells folded inward to form the retina. The skin on the surface became transparent to function as the lens that focused light on the retina. These were housed in cup-like depressions that enable light-sensitive cells to delineate the direction of light. Today, at one end of the spectrum you have coelenterates and ctenophores like jellyfish with pigment eyes, pigment cells tied to sensory cells and covered with a cuticle that acts as the lens. Worms, insects and mollusks, with varying degrees of complexity, have these kinds of eyes. In the animal kingdom as a whole, you have two kinds of image-forming eyes, simple and compound. Most arthropods, such as shrimps and insects, have compound eyes. Hundreds of optical units called ommatidia form separate images on retinal cells creating a mosaic and sometimes a combined image. Vertebrates have simple eyes that are structurally similar to the human eye; there is an increase in complexity as you ascend from fishes to amphibia, reptiles and birds. Some nocturnal animals only have rod cells, but these are greater in number and sensitivity than in humans, enabling them to see better in the dark. Taxonomists postulate that eyes evolved independently at least forty times in the animal world.

Guru: Let's consider the evolution objection first. As a body of theory, evolution is an elegant paean to the unity of all life. But the scientific claims of various evolutionary theories are not to be confused with the metaphysical pretensions of materialists. While a given evolutionary theory might paint a plausible picture in explaining a particular phenomenon, it's simply superstitious to assume that that theory is a key opening every door.

In talking of the physiological structure of the eye, I had no intention of introducing a *deus ex machina*. Rather, I wanted to outline the hard facts we encounter today so as to move towards an explanation that makes sense of them. But before going there, let me address the status of the eye in the context of evolutionary theory.

From the time of Darwin, the eye has posed problems for most evolutionary theories. At least on the face of it, the assertion that a light-sensitive spot eventually became the complex organ that is the human eye is hardly believable for two reasons:

1. A complex organ that inherently depends on a multitude of other entities to function at all could not evolve gradually. If vision is to work, we need, all at once, the eyeball with its settings, the focusing mechanisms, the connections to the nerve cells, and the brain with its capacity to see. The eye couldn't evolve on one path, the nervous system on another and the brain on a third – it's all or nothing. Where the eye appears in the fossil record, it appears fully formed, and there's no clear account of the evolution of the components of the eye, such as the retina with its millions of neurons or the lens that focuses light. The nautilus today has the same pinhole eye it had hundreds of millions of years ago with no apparent change.

2. The notion that there was a progressive increase in complexity in a primitive seeing organ is difficult to defend because the different kinds of eyes are structurally so different that evolutionary scientists like George Gaylord Simpson and Richard Dawkins do not claim they evolved from each other. In fact, as you mentioned, they even talk of the eye having evolved independently many different times through, as Dawkins put it, the power of natural selection.

Geek: What you have said of the intricacy of the human eye is true enough. But there are two other sets of inconvenient facts that stand in your way:

1. Darwin's idea that the complicated systems we see in nature emerged through thousands of small steps over hundreds of thousands or millions of years has been validated on various fronts, and this includes the eye. In

"A Pessimistic Estimate of the Time Required for an Eye to Evolve," Dan Nilsson and Suzanne Pilger have given a very plausible scenario, with very few mutations involved, for the evolution of the fish eye from an eyespot over a period of 350,000 years. The sequence proceeds from an eyespot on the surface to a depression that increases acuity (flatworms) to an aperture with jelly inside (the nautilus) to the formation via mutation of a transparent layer that becomes the lens. Brains actually came after eyes, and the claim that you need everything at once is simply not true across the board. Eyespots came first, the wiring of the nervous system second. Eventually, they evolved in parallel, with nervous systems gradually improving to cope with greater sensorial input. The fish is a great illustration of this because it has an optically excellent eye but hardly any brain. No images are formed of what it sees. And we are not talking here of half an eye eventually becoming a complete eye but of a gradual increase in the quality of vision.

2. Those who believe that a deity designed the processes in nature turn a blind eye (!) to the many imperfections we find there. The human eye is a prime example. The retina is not functionally efficient because it points backwards. The photoreceptors in the retina face away from the source of light. When a ray of light comes in, it goes through myriad neurons and capillaries before reaching the receptors that read them. Once light reaches them, the photoreceptors send signals back up the retina to the optic nerve. The optic nerve connects to the brain through a hole in the retina called the blind spot. Two consequences of this construction are evident: the light that reaches the photoreceptors is degraded because it has to pass through various layers, and the blind spot is blind because light is blocked at the site of the optic nerve. In the eyes of squids and other mollusks, on the other hand, the light-sensitive cells are pointed towards the light and signals are sent from these nerve-endings right to the brain. These defects in the human eye can be explained

in the context of the evolutionary process, but make no sense if we have to invoke a Creator.

Guru: This is another instance of looking at the same facts and drawing different conclusions. Let me start with your second point. The degradation of light caused by the neurons and blood vessels has no ultimate impact on our vision because the eyes move constantly to change the shading around the objects we see, and the brain generates a clear image from the many shaded versions. Likewise, the effects of the blind spots are cancelled out because we have two eyes. Each eye is blind to a different part of the visual field and the brain combines both views to give the complete image. Most important, the positioning of the capillaries and epithelial cells behind the retina actually makes the eye more efficient. The retinal cells get their energy from the capillaries and regeneration from the epithelial cells. If the retina were to face forward, capillaries and epithelial cells would block the light it receives, making it less efficient.

With regard to the elegance of the eye's engineering design, let me ask you first whether it does its job well. You step out on a balcony and see hundreds of buildings and the landscape as it stretches many miles into the distance. Although your retina is one centimeter square in size, it's able to tell you that your field of vision covers a vast area and not a tiny speck of an image. The light that enters your eye is guided through this one and only transparent part of your body to the photoreceptors at the back of the retina, where it triggers a series of chemical reactions that go back up the retina and through the optic nerve into the brain. Despite the alleged shortcomings you highlighted, it is patently obvious that:

♦ the entire complex process takes place all-but instantaneously without your being aware of it

♦ what you get at the end are accurate images of the world around you.

We have not yet considered evolution, but at this stage it's worth observing that the paradigm of infinite Intelligence expressly assumes that the finite is not only less than perfect but also subject to certain processes and laws of development and growth. The crudely anthropomorphic super-tinkerer is a straw god. Infinite Intelligence, on the other hand, manifests itself through the creativity and cooperation, the rationality and contingency that we find in nature.

The next objection was the idea of intermediate forms in the evolution of the eye. About the Nilsson-Pilger paper, David Berlinksi points out that, as confirmed with him by Nilsson, the authors *do not* offer a computer simulation of the evolution of a functioning eye from light-sensitive cells driven simply by natural selection and random variation. What Nilsson-Pilger sought to show, says Berlinksi, was the evolution of an eyeball and not an eye using "back-of-the envelope calculations." Despite all the research into mammalian visual systems of the last twenty years, he concludes, "no theory has anything whatsoever of interest to say about the fact that the visual system terminates in a visual *experience*, an episode of consciousness. We cannot characterize the most obvious fact about sight—that it involves seeing something."

Standard accounts posit that the phyla diverged over 500 million years ago, and that each organ is the result of thousands of random molecular genetic mutations. How is it possible, we might ask, that five phyla, which each took separate evolutionary paths, with thousands of random mutations, somehow ended up producing similar kinds of eyes? More to the point, how can you propose intermediate forms while also holding, along with other evolutionists, that the eye has evolved separately at least forty times in different life-forms?

> *Geek*: I know I said that, but you should know that recent findings also show that one and the same gene group, Pax-6, and its analogs play a key role in the development of the eye in the five genetically different phyla with visual systems. In this context, at least, there is no contradiction in also holding that the eye is subject to evolutionary pressures and adaptations.

Guru: Surely you realize that these findings are especially significant for those who hold as I do that modern science bears witness to the wonder of the world. In his trilogy on science and God, Gerald Schroeder points out that the idea that a single genetic blueprint governs the development of this delicate and intricately organized organ in the most varied species is an indicator of underlying rationality rather than sheer randomness. Not only do the Hox genes and master-control genes like Pax-6 build each organ, says Schroeder, but identical versions of these genes give different instructions in different animal bodies. Thus, flies with their compound eyes have the same Pax-6 gene as humans. Also, fully functional eyes appear right at the beginning of the Cambrian explosion.

In the light of this genetic data, many evolutionary theorists have given up the idea that the eye evolved separately in the five phyla. Stephen Jay Gould has said that there is no longer any separate story of evolutionary origins for the eye, because underlying genes common to all phyla positioned the fundamental structure of the eye on one and the same genetic pathway. In his final *The Structure of Evolutionary Theory,* he points out, "We must now grant strong probability to the proposition that, absent an 'internal' direction supplied by the preexisting Pax-6 developmental channel, natural selection could not have crafted such exquisitely similar, and beautifully adapted, final products from scratch, and purely 'from the outside.'" Theorists now hold that the eye appeared from a common genetic tool-kit that made the Cambrian explosion possible. These theorists insist that the tool-kit came from a common ancestor while also admitting that there is no fossil evidence for this ancestor. *A more plausible conclusion is that the Pax-6 gene, like other master-control genes, was pre-programmed into the very stuff of being.*

So much for the appeal to evolution. But my point was far more fundamental. Even if we grant the evolutionary account of the origin of the human eye, we are left with the two other questions I asked earlier. Why is reality structured so that there can be a symbiosis between light and the eye, between photons and neurons, that results in sight? Light waves, incidentally, are the only kind of electromagnetic radiation with the energy level required for biological systems to detect them, and the wavelength of light is just the right size for the high-resolution camera-type vertebrate eye. Secondly, what lies at the basis of the amazing transformational interfaces between eye, brain and mind that convert electromagnetic waves into electrical signals and then into subjective visual images, activities that take place so effortlessly that we take them for granted? And let me re-phrase my first question about the origin of the eye to ask: how and why did the most primitive light-sensitive cells come into being?

> *Geek*: Your focus seems to have shifted to the origin of light-sensitive cells. The chemistry of light-sensitive cells, embodied in processes like photosynthesis, was present well before eyespots started appearing. Since the chemical switches required for light-sensitive

cells were available a billion years ago and since eyes are useful for an organism's survival, why is there anything remarkable about the emergence of vision as such?

Guru: This misses my point entirely. Can dust and debris magically develop light-sensitive cells given a hundred trillion years—or ever? And how can a collection of chemicals work together at random to generate a system that:

- ◆ processes the equivalent of a thousand million computer instructions at every moment
- ◆ produces, through chemical amplifiers, up to 100,000 messenger molecules from a single photon
- ◆ constantly creates images of the world?

This is the physiological bottom-line that, I maintain, is inexplicable if there is no underlying creative intelligence at work. But beyond the biological, I must draw your attention to the ontological hard facts. As I mentioned before, ontology is a description of the structure of reality that cannot be proved or disproved by science but is presupposed by it. An example of ontology is the insight that a valid inference from a valid premise leads to a valid conclusion. This can't be proved by science.

The physiological transactions we've reviewed are stepping stones to the ontological hard fact that is the central mystery in the phenomenon of physical vision: the fact that you and I can become directly and immediately aware of the world around us. Why is it the case that one piece of matter attains the capacity to become aware of another in so intimate and complete a fashion? Can anything in physical reality explain or even describe this ability to *be aware*? Viewed as a mode of awareness, the phenomenon of seeing is irreducible to anything physical or material.

> *Geek*: Now you've lost me. To start with, you postulated a higher intelligence as the *deus ex machina* that explains the evolution of the eye. Now you seem to be saying that each one of us is a mini-*deus,* if I may stretch the expression!

Guru: Infinite Intelligence is not just "higher" but the highest possible intelligence. By now you should know that I'm not touting either a *deus ex machina* or a god of the gaps; this is clear in my discussions of the existence of God. By talking of awareness as inexplicable in physical terms, I'm not suggesting that we're mini-gods, whatever that means. Rather, I'm drawing your attention to something obvious in our experience that is rarely realized or even considered. No evolutionary biologist has given an explanation for seeing in terms of the phenomenon of awareness. In fact, the thought seems hardly to have struck any of them. So let me bring this issue back to the forefront where it belongs.

At this point I'm not considering the processes of consciousness or thought, both of which we will discuss down the road. Neither am I really addressing the nature of sense-data, a topic that has engaged philosophers for many years. My concern is this basic ability to be aware. Awareness as exemplified in seeing is a new kind of reality that transcends the physical and is irreducible to anything else. What do I mean by this? Well, what happens when we see? The objects in our visual field in a certain sense become a part of us, not physically of course, but mentally. Whether it's a mountain range or a shopping mall or a friend, we become aware of them directly and without mediation to the point that they enter our minds. This is what it means to see.

Now none of this could happen if we didn't have our mysterious organ of vision and its equally mysterious interplay with light. But the end-result we experience as visual awareness is something that could neither be predicted from the physical components that make it possible nor be described in physical terms. To see is to be aware of our environs in the most intimate fashion. All of our sensory organs enable different modes of awareness of our surroundings, but seeing is the most dramatic and direct of these modes. Seeing makes the world real to us. Although the things we experience are not simply collections of our sense-impressions, the visual and other kinds of sensory data they induce in us are pre-requisites for most thought and action.

It is this dimension of seeing, its introduction of the entirely new reality of awareness, that makes it a marvel and a mystery of the first order. We know that the images we see on a TV screen are not entirely reducible to cathode ray tubes and phosphor-coated screens. These components simply act as a medium to receive and display the images. Analogously, we can think of the structure of the eye and the constitution of light as conduits for a new dimension of the world that is qualitatively different from the physical. To send a nerve signal to the brain is one kind of thing; to be aware of the cars passing in front of us by opening our eyes is another.

Coming full circle, I submit that all dimensions of seeing, from the existence of light-sensitive cells to the interaction of photons and neurons to the reality of visual awareness, are inexplicable in terms of a purely material matrix.

> *Geek*: What you're saying is that there's more to seeing than simply meets the eye! You must know, however, that what you call awareness has been the subject of new research programs on visual consciousness run by a whole slew of scientists. So they're not oblivious to the issues you raise.

Guru: I did point out that my concern at the moment is not with consciousness as such. That's something on which I have much to say. I've simply drawn your attention to the indisputable fact that seeing at bottom is becoming aware. To talk of one piece of matter becoming visually aware of another is to use language and think thoughts that simply don't fit into physics or biology or even optics for that matter. It is nevertheless a reality we experience every time we open our eyes. All of the debates about the origin of the eye seem to me just squabbles over whether the components of the TV set show marks of a creative intelligence. We've missed the big picture that includes not just the components of the set—these are, of course, important in themselves—but also the transmission and reception of images, a reality not reducible simply to the components.

And as if seeing, understood as awareness of the physical, were not mysterious enough, we enter an even more mystifying dimension when we migrate from seeing with the brain to seeing with the mind.

> *Geek*: I think you've built a fairly good case up to this point. But I don't quite get your distinction between seeing with the brain and the mind. And I obviously need evidence, tangible, scientifically demonstrable evidence, to take it seriously.

Guru: I agree on the need for evidence. But the criteria of evidence must be carefully thought through. If I'm making a scientific claim then I will need scientifically demonstrable evidence to support it. If I'm making an ontological claim, and this is the kind of claim I'm making in this discussion, I must support it with undeniable ontological principles or hard facts of experience.

Chapter

Seeing with the Mind

GEEK: Let's do a reset. What conclusion are you driving at here, or, more to the point, what's your point?

Guru: My point is the wonder of the world. If we recognize the truly amazing nature of sight, then our minds will be better equipped to apprehend, appreciate and assimilate the glory of being. If you're blind to sight, then it's hard for you to see anything else!

Let me briefly review where we've been so far and then introduce the idea of seeing with the mind.

The more we learn about the mechanism of vision the more we marvel at its very existence. It is difficult to call the phenomenon of sight anything but a miracle. If it originated in light-sensitive cells, how did it come to be the case that reality is structured to allow a symbiosis between light energy and the physico-chemical structure of brain and eye? Did the eye develop in order for us to see, and thus for a purpose? Or is it simply an accident, natural selection working its way from the Big Bang through the Cambrian Explosion hand in glove with random mutations? But even if it were beneficial for survival and even if random mutations independently produced the same results in five phyla, vision could still not exist if certain underlying structures did not make it possible.

Of course, an obsession with speculation about the past can easily blind us to the present. Beyond all the historical details, consider our own experience of seeing. On the one hand there are the millions of neurons at work, the iris adjusting the pupil, the lens thinning or thickening to focus light on the retina and information being processed faster than in any supercomputer at the bat of an eyelid. On the other, we directly and constantly experience all that we see in our minds. The act of seeing allows us to get "in touch" with things. To see is to be aware and thereby to know what's going on in the world around us. Acts of awareness are not simple optical processes. Optical processes enable us to be aware, but to identify being aware with the operation of optical processes is like saying that the images on TV screens are to be identified entirely with electron guns and screens. The enabling structure is one thing and the enabled experience another. Awareness is an ontological reality as fundamental as matter itself.

But so far our discussions have remained at the level of physical sight. Let me pass on to the equally important theme of seeing on a purely mental plane. Here we are talking of the mind's capacity to grasp meaning that is the exact counterpart on the conceptual plane of what takes place during the act of seeing at the perceptual level. When we "see the point" or "see something to be the case" or "know what you mean" or "realize" or "understand" or "comprehend" or "visualize," then we are performing acts that, again, cannot be described or explained in physical terms, let alone by reference to evolution. All of these acts presuppose the mind coming to grips with something. When we say that a sense of integrity should prohibit us from taking a bribe, we're not talking simply of physical actions but of irreducible ideas. When our reading of Mahatma Gandhi's *My Experiments with Truth* moves us, it's not the print marks on paper that stir our emotions. Our mind sees something that can't be expressed in terms of molecules or particles.

No metaphor better describes *the mind's act of grasping meaning* than that of visual perception. A few examples will illustrate this. Two people who look at the same body of data can see different things depending on whether or not they have better "vision." A botanist will "see" more in a garden than someone who's unfamiliar with plant life, although they're both looking at the exact same body of plants. A poem in Japanese will seem to be simply print marks on paper to someone who knows only English, but will be read and appreciated as an instance of haiku by a native

Japanese speaker. In this last example, we could talk of the observers seeing one and the same object *as* two different things. Again, just as we can see something in a room without noticing it, we can likewise be familiar with some particular fact without grasping its significance.

The metaphor is just as applicable in a converse context. If we have defective vision, myopia or congenital blindness to take two instances, our ability to see clearly or at all will be restricted by our particular condition. On the level of human knowledge too, it is possible for one person to be blind to what the other sees to be the case. There is, of course, the possibility that the person who claims to see something may just be hallucinating. But there are nevertheless ways in which we can discriminate between fact and fantasy, illusion and reality.

> *Geek*: I can accept what you say about our ability to grasp meaning without necessarily jumping to your conclusions. After all, animals do this, as do computers. The dog knows where it buried its bone. Computers can beat humans at chess. You may be too hasty in writing off an evolutionary explanation based on physical mechanisms. Francis Crick has done some interesting work here. Meaning arises from correlated firings of neurons and linkages to related representations, as he puts it.

Guru: We'll get to Crick when we consider consciousness, but it should be obvious to you that the neuronal activity that accompanies your act of seeing the meaning of "Give me liberty or give me death" is not the same as understanding what this thought implies.

Let me show you what I'm saying just by pointing out what you're trying to do. You're trying to convince me that no independent meaning-processing takes place in our minds. This is what you *mean* to communicate to me, this is the point you want me to *see*. But is your point, the meaning you wish to communicate, purely a matter of neuronal transactions and brain states? Are you trying to alter certain of my brain events or to get me to see something to be the case? If it's the latter, then meaning is something different from physiological states of affairs. You could say it's dependent on a physiological substrate, but you would still have to admit that it's not the same thing as the physiological transaction itself.

Geek: Of course, our experience of meaning seems different from the neuronal firings that express it, but this is just a matter of appearances. Fundamentally, we have no reason to believe they're not one and the same thing, and there's no way to show that they're different.

Guru: Well, we will be discussing the mind, and I hope to show then that this sort of a materialist position is plainly false. But at this point I'm trying to simply get you to see that your mind does see certain things to be the case. Whether or not your experience of seeing something to be the case is actually just a series of brain events is a point to be taken up separately. But you do agree, don't you, that you've sometimes seen something to be the case? Or do you hold that you have seen it to be the case that you can't see anything to be the case?

Geek: Very droll. OK, I do sometimes see things to be the case. But that's not saying much. I'm simply affirming that some of my current neuronal firings tie into the brain's associational database. No invocations of supernatural entities are required.

Guru: But it's not as tidy as you represent it. When you see something to be the case, you do so because you have one or more reasons for believing this to hold. After all, you are the one who asked for evidence. You can't be convinced of something by the action of certain physical causes in your central nervous system. You're convinced by reasons. You disagree with me because you haven't seen enough reason to agree. Meaning is all about reasons and not causes. If I say a poem is beautiful, neither my message nor its truth is reducible to neuronal excitement in given regions of the cerebral cortex. It's all about concepts, reasons and meanings, not causes and effects. I cannot see how you can dispute any of this without blatant self-contradiction.

Geek: What about my analogies of dogs and computers? When Jackie hurls a Frisbee, Fido races to fetch it. Fido knows that this is what Jackie wants him to do and acts accordingly. And when Big Blue beat the world chess champion, the computer had an objective it achieved by deploying its problem-solving skills. If dogs and computers can see meaning, what's so special about this capability?

Guru: Simply this. The higher animals certainly have some degree of consciousness but it's restricted to perceptual activity. In other words, they respond to certain perceptual objects when these are displayed, but they're incapable of pure conceptual thought, particularly when there's no perceptual component to the activity. Certainly, there's no evidence that animals can deal with concepts that lie outside sensory experience, such as mathematical abstractions or concepts of justice and beauty. But seeing with the mind is a singularly conceptual activity.

The difference between the computer and the mind is this. The computer isn't conscious or aware of itself doing its processing as we are when we think. Rule-governed activities, such as chess, can be formalized and simulated in a computer program. But remember, it is we humans who do the programming and create and input the appropriate symbols. The computer has no clue about the meaning of the symbols it processes. All that it does is generate electrical pulses. It doesn't know what it's doing. There is, in fact, no "it" to know because it's simply an assemblage of mechanical systems. There's no knowing going on within it; there's simply a continuous flow of electrical pulses through its circuitry, assuming, of course, that it's connected to a power source and appropriately programmed by a user. Real thinking, on the other hand, involves our knowing what we're thinking about, being aware that we're thinking, recognizing that we're arguing or reaching a conclusion and the like. While both computers and minds store immense volumes of data, the computer's processing of the data is the manipulation of electrical pulses on pre-programmed paths. The mind, on the other hand, comprehends the data, *knows* its *meaning, uses* it. This is what seeing with the mind is all about. It's something we do instinctively during most of our conscious life.

Geek: OK, I grant that we do see meaning. But I'm still not convinced that this is anything other than an act of the brain. What need is there to go beyond the physical?

Guru: For one thing, the two of us are communicating using language, a remarkable phenomenon, involving as it does the complexity of syntax, that we overlook simply because we do it all the time. I've noted that he process of linguistic understanding does not have any neural correlate and is not the activity of a specific bodily organ. Words are symbols or codes signifying something, and the coding and decoding activities required in using language presuppose an entity that can endow and perceive meaning in symbols. Can a material object perceive meaning? By its very nature, the act of comprehending the meaning of something is non-physical. Moreover, many of the words we commonly use designate things that either have not been perceived, e.g., quarks, or cannot be perceived by the senses, e.g., justice, but they have meaning. The acts of giving meaning or seeing meaning are again irreducibly immaterial and are carried on all the time as illustrated by our use of language. And I'm sure you'll agree that this meaning-driven activity we call language is by no means rare.

The upshot of all I've said is that, like physical seeing, the seeing of meaning is an ontologically new reality.

PART II
The Invention of Nature

*I*N MY VIEW, our discussion about seeing shows that there is at least this one phenomenon that cannot be explained simply by reference to the world of matter. But you know that my overall thrust is more fundamental and far-reaching than just visual perception. I hold that literally everything in the world, the world included, which is nothing beyond the collection of things that comprise it, is precisely as mysterious as seeing. The human mind instinctively seeks an explanation for every thing, being, activity and phenomenon in the world. At one level, such explanations are found in describing the relationships between things, and these are the laws of science. But beyond scientific explanations, we seek ultimate explanations that tell us how there could be anything at all and why the world is constructed in such a way as to follow incredibly intelligent laws. I have proposed that the only satisfactory ultimate explanation is an infinite Intelligence that brought all things into being. But before we can see this to be the case, we should first recognize nature, by which I mean all natural systems, for what it is: an astonishing invention.

Geek: A fascinating idea. But again, from my standpoint, it's one that needs to be substantiated. Why do we need an ultimate explanation beyond the scientific one? And what does it mean to call nature an invention? You've certainly gone beyond science, but in the direction of poetry. Rich in metaphor but light in substance.

Guru: I've already laid out the distinctions between scientific and meta-scientific questions. But here you're entitled to an explanation of what I mean by calling nature an invention.

Who Holds the Patent on Quantum Fields and the Genetic Code?

The world is made up of things: stones, sunsets, flowers, tornadoes, microbes, chimpanzees, and galaxies. The things in turn have properties of various kinds: from color, size and shape to sound, smell and taste to locomotion and replication.

The most obvious and yet the least noticed feature of these things is their individual uniqueness and their inexplicable ability to follow certain laws. At one time there was nothing, or a formless something if you prefer, and then the world as we know it came to be with a plethora of particular things that obey certain universal instructions.

Now someone invented everything around us in the human world: cars, cookies, restaurants, microchips, and clothes. As we noted, all these things were first ideas in the minds of their inventors before they were brought into being. When we turn to nature, we ask who or what thought up photons and suns, DNA and dinosaurs, mass and charge? Who or what is powering the whole enterprise, keeping it always "online?"

We call something an invention because it's the concrete actualization of a unique idea. Inventions embody ingenuity. The systems in nature and the laws of nature are far more complex, intricate and innovative than the internal combustion engine or the submarine. In fact, human inventions are impossible without the existence of natural laws and systems. Nature with its systems and laws embodies ingenuity, uniqueness, and plenitude. To the extent that it embodies an Idea, it too is an invention.

Of course we can give the physical, chemical and, where relevant, biological conditions that gave rise to every one of the things in nature. And, of course, we can delve into their subatomic antecedents so as to thoroughly chronicle and predict their wavefunctions and world-lines. And we can further classify them under different categories: phylum, genus and species; igneous, sedimentary and metamorphic rock forms; infrared, ultraviolet, gamma and X ray radiation; quarks and leptons, and the like. And to terminate further inquiry we can take final refuge in the cosmic cauldron that was the Big Bang, the primordial ooze of prebiotic evolution and the fiery hand of natural selection.

But the fundamental dimension we are concerned with here goes beyond any description of the development of a thing and its classification under some scheme. Our focus is on the blueprint of its being. If a set of primordial states and initial conditions gradually unfolded to produce a given thing or being, these states and conditions were programmed at the start to produce this end-result. *Who or what wrote these programs? Who holds the patent on the process and its products?* If we ask why grass is green, our question is not about chlorophyll or molecular structure or energy sources. Given that certain ultimate physical processes gave rise to a blade of grass, then its greenness and other properties were contained in those processes from the very beginning. Who invented the greenness, or for that matter the blade of grass, which were, so to speak, thus implanted in the fabric of physical reality?

There's no escaping this question by attributing it to the "genius" of evolution. Let us grant this appeal to evolution and theorize that all things emerged through processes of evolution driven by some mysterious mechanism. But our question is not about the particular pathway taken by the raw materials of the physical world as they proceeded toward the present day, which is the eminent domain of any theory of evolution. Rather, our inquiry concerns the inherent capability and potentiality of these raw materials to end up where they are today. The multifarious manifestations of mass/energy that we encounter today would not have been possible if mass/energy did not start off with the acorns that would later become oaks. Moreover, there would also have to be a systematic set of laws governing all things. How were these laws conceived, and just as important, how were things forced to follow the laws? What compels the protons and neutrons in an atom to be bound by strong nuclear forces that are strong enough to overcome the positive charges that could blow apart the nucleus? How is it that mass-energy is always conserved?

On any account of the world, then, we have to acknowledge that there's a blueprint underlying the existence of all the things that constitute it. Moreover, it's an inventor's blueprint because each thing with its particular conjunction of properties and its ability to follow certain specific laws is framed in terms of an idea, and it's an idea that comes to life. Every petal, every quark, every molecule of gas was thought up before it was materialized.

There's yet another level at which the world may be thought of as an invention, and that's simply the very fact it exists. We start with the hard fact that the universe just happens to be here, when it might very well not have been since there's no logical necessity that it should exist, and that we don't know how it got here. We might have theories about vacuum fluctuations or big bangs but we have no idea how, to begin with, any of this happened to be around, why there were laws that governed all things at all times. Atheist scientists like Steven Weinberg admit with barely concealed embarrassment that the initial conditions responsible for the formation of the universe are just "given," which is just a way of saying this is the way it happened to be and we don't know how it got to be that way. As a matter of fact, it is not just these initial conditions but all subsequent events in the history of the universe that are given, since there's no reason why one event should cause another. The universe itself is the "given." But given by whom?

Every atheist says that the given is simply a "brute fact." We should accept it as "just there." But why should I stop here? Why cannot I wonder how a reality of such depth and diversity came to be? *If it was brought into being then it was invented, for to invent something is to make or design something that didn't exist before.* Thus the universe is either entirely inexplicable or it's an invention. The rational mind can't be satisfied with the idea that the world is simply a brute fact. The scientist can't consistently believe that everything has an explanation, but that the chain of explanation begins and ends with facts that can't be explained. In addition to its very existence, the universe discovered by science bears all the hallmarks of any invention: creativity, innovation, a unifying theme coupled with diversity of expression. And it is clearly the invention of an Artist!

As I've tried to show, great scientific minds, from Isaac Newton to Albert Einstein to Stephen Hawking, have in fact recognized that the universe is indeed an invention, the invention of infinite Intelligence. As these scientists see it, science becomes especially exciting if we recognize its true mission as the study of an invention generated by an infinite Mind. Einstein said that it is the cosmic religious experience that is the strongest driving force behind scientific research.

To grasp the inventedness of the world, we need to both reflect on the big picture and look at the universe as if we were seeing it for the first time. Now you've described this as a poetic vision pre-dating science. Let me remind you that many of the greatest modern scientists were driven more by the beauty they discovered in the structure of the world than by any other consideration. Richard Feynman, for instance, said that truth can be recognized by its beauty and simplicity. The poets have much to offer as well. William Blake talked of seeing a world in a grain of sand. Gerard Manley Hopkins sought the dearest freshness "deep down things." These may have been pre-scientific intuitions, but they seem to me perfectly compatible with all that we've learnt about fields, particles and space-time.

> *Geek:* That was an eloquent defense. But I still say you can't use poems as arguments or prayers as proofs. Sure, you can see the world *as* an invention of a deity. But the question really is whether it *is* indeed such an invention. You haven't offered any evidence to support your assertions of a celestial Patent Office. All so-called beauty in nature can be seen simply as a manifestation of natural selection at work. To use Darwin's example, flowers have brilliant colors in contrast to surrounding green foliage so that they can attract the attention of insects.

Guru: This was by no means intended to be an evidentiary exercise. I was trying to give you an idea of where I was going while setting the stage for my next exposition. I want to emphasize again, however, that modern scientists from Einstein and Planck to Schrödinger and Heisenberg were led to transcendent conclusions through their investigation of the natural world. I will focus on trying to lead us down the trail they blazed. With regard to the Darwin reference, let me tell you that there's still the question of who or what pre-programmed the colors and scents of flowers into the fabric of reality. I'll try to show you how this ultimate question keeps popping up virtually anywhere you look on the landscape of modern science.

> *Geek:* OK. So how do you propose to start?

ℭℐ

The IQ of the Universe

Guru: Let me put my cards on the table here. Like the sudden appearance and outward expansion of the universe itself, the genesis and growth of life on earth seem to me unmistakable manifestations of extraordinary intelligence, beauty and power. There is a steady progression of foundational transformations that proceed in this sequence:

1. An initial matrix of unutterably intelligent laws

2. The application of these laws in precisely the direction that maximizes the IQ of the whole system so to speak, despite the most inhospitable conditions imaginable from beginning to end

3. The manifestation of a hierarchy of radically different kinds of intelligence: matter, life, innately purposive structures, consciousness, and mind.

What strikes me about this whole dynamic is the incredible complexity that exists from the start and the systematic appearance of new kinds of things (life, consciousness) that cannot be put together from whatever existed before.

The old Anthropic Principle would say that the emergence of life in the face of wild improbabilities of every kind seems to show that the universe or Someone behind the scenes wanted life to appear and prepared conditions that were amenable to its arrival. To which the atheist says there are alternative explanations that are equally satisfactory, that of pure chance or of an infinite number of universes in which every possible history will be realized.

But the Intelligence Principle, if I may call it that, starts with the claims that (a) any appearance of intelligence requires an intelligent source, (b) the history of the universe shows intelligence manifesting itself spectacularly in successive stages, and (c) the culmination of this process is the emergence of self-conscious intelligence. I certainly accept the possibility that chance on its own could produce certain primitive kinds of order, but I consider it obvious that chance cannot be the source of precise, intricate, inter-connected laws. A multiverse scenario requires laws governing the ensemble of universes, so we are still left with the phenomenon of intelligence at work. The multiverse maneuver is in any case a desperate scheme adopted by those who find chance implausible but cannot stomach the idea of a creative Intelligence. Self-organization and complexity are other alternatives, but these likewise have to be grounded in an Ultimate Source. At the end of the day, we need a credible explanation of the data at hand. To my mind the only workable model is the idea that the ultimate origin of matter, life and mind lies in an infinite Mind.

In making the case for intelligence as the fundamental reality, my road-map proceeds from the macroverse of space and time to the microverse of the subatomic world and from there to the origin of the universe and of life, consciousness and mind. In each of these areas I consider established and currently popular theories.

My overview of the foundations and frontiers of science culminates in a study of the laws of nature that prominent scientists have termed the Mind of God. Modern science is made up of theories that seek to describe different dimensions of the physical world in terms of laws. Each theory has to make sense of the observed facts in its domain and also fit in with accepted theories in overlapping domains. Thus theories about the behavior of quantum particles have to cohere with theories about the behavior of stars, since a star's activities are driven by the properties of the particles that comprise it. Thus the great theories of modern science from relativity to quantum physics to DNA seek an integrated, coherent picture of the genesis and behavior of matter, energy, motion and life.

Every scientific theory is in principle revisable. At the same time, it would be safe to say that theories and their extensions are made up of claims that carry differing degrees of probability depending on the evidence in their favor. In order to maintain clarity and make progress in our investigation of the world revealed by science, we have to make distinctions based on levels of evidence. Given this starting-point, we will evaluate theories based on an evidence-based gradation system:

1. hard facts established beyond reasonable doubt (an A grade)

2. reasonable hypotheses based on circumstantial evidence (B)

3. speculation without any physical evidence but with some degree of plausibility (D) and

4. speculation that lacks evidence and flies in the face of what we know to be the case (F)

If you crack open any recent book on the state of physics and cosmology, you are likely to be bombarded by a welter of complex equations, esoteric concepts and intrepid speculation. As we shall see, the correlation between mathematical theory and the natural world tells us a lot about the intelligence embedded in fundamental structures. In the context of scientific theory, however, we should be aware that a mathematical model cannot establish the validity of a hypothesis in the absence of empirical, tangible evidence. The final arbiter of any scientific theory is experimental evidence. Mathematics is the cart that draws a theory, but evidence is the horse that draws the cart. We must beware of what John Maynard Smith called "fact-free science," referring to Stuart Kaufmann's self-organization scenarios.

Finally, we've seen that there are some questions that, by their nature, cannot be addressed with the methods of science. These are the questions of ontology and, as mentioned, much of the confusion in popular science comes from a tendency to confuse scientific and ontological questions.

> *Geek*: So what you're proposing is a pretty comprehensive survey of contemporary science. To make things flow smoothly, I'll reserve my comments and queries for the end of each part of your presentation.

Guru: I appreciate that since I'm going to be sending you one chapter at a time of my infamous manuscript! Having said that, I want you to feel free to cut and paste as it makes sense. If any part of a presentation raises concerns, just cut it out and send it back with your comments. This is supposed to be a living document, at least for our present discussion.

Chapter

Space and Time

S CIENCE IS DRIVEN not just by questions of "What" and "Why" but "Where" (Space) and "When" (Time). Where and When are related to What, Why and Who. Small mistakes about Space and Time can trigger off monumental mistakes about the origin and nature of the world. Since confusion has been more the rule than the exception here, this will be an appropriate starting-point for the next stage of our investigation.

So what is space? And what is time? And how did they originate, and when and where? Einstein is said to have remarked that space is what you measure with a ruler and time is what you measure with a clock. But we still want to ask if space is something absolute and infinite. Does it stretch on and on without end? Is there any mapping system that locates specific points in space the way that latitude and longitude identify each point on the globe? Likewise, is time something absolute? Is there a cosmic clock that keeps time for the whole universe or a Greenwich Mean Time against which all other times in all the galaxies are measured? And when did time begin, and if it had a beginning what happened "before?"

I want to start by making what sounds like an extraordinary claim: *There is no such thing as space that exists over and above and alongside all the things in the universe.* And *time is not a flowing river that exists independently of events and agents that cause events.* These insights are neither new nor implausible.

They date back to the ancient Greeks and have been brought to light again in modern science, most especially in Quantum Gravity. As you've been the first to say, science has shown that many commonsense views of things are simply mistaken. We think that the sun rises from the east and sets in the west, but, of course, it doesn't. The world looks flat, but it isn't. On a more fundamental level, Einstein's Theories of Relativity have excised the old ideas of an absolute space that contains all the things in the world and an absolute time that exists separate from things and events. Newton's container view of space and time has been replaced by a theory of space-time in which space and time are married to matter and motion. But this corrective surgery is still a work in progress.

In many discussions, space, time and space-time have taken on a life of their own with elaborate properties powered by complex mathematical formulations. All sorts of people, scientists, mathematicians, philosophers, and journalists, have said all sorts of things about these matters. But, at the end of the day, we have to differentiate between hard facts and sheer speculation. To get to what is real and fundamental, we have to exorcise the merely speculative from our minds while grasping what is obvious and undeniable.

So what exists? At a fundamental level there is certainly a succession of particle-events related to each other, of effects caused by various agents. At a higher level, there are processes and structures and relations between these, i.e. laws. Beyond all this, we're aware that we exist as persons who intend to bring about certain events and sometimes act on our intentions. At another level, we know that matter exists and we know that all matter in our experience is in motion. We know that we exist and that we're not just matter, a point I will try to elucidate as we proceed. Now science generates laws and theories that document the properties of matter and motion. Some of these properties are impossible to visualize and can only be described in the refined language of mathematics. Nevertheless, the most ingenious mathematical solutions are superfluous if they cannot be verified in the physical world.

To be honest, almost everything exciting in modern science is simply speculation, such as M–Theory, inflationary theory and the like. And to be sure, many established theories started out as just speculation. But let's never lose sight of the difference between conjecture and data. We have to draw a line in the sand between interesting ideas and hard facts established by experiment. By all means, let's continue to speculate and brainstorm and ruminate. But let's never fool ourselves as to what's fact and what's not.

If sheer speculation is one end of the spectrum, sheer confusion is the other. Many scientific discussions are marred by elementary errors and fundamental fallacies from the standpoint of rational thought. For instance, the assumption that nothing is a kind of something has wreaked havoc all across the board. A vacuum is a "something" with a structure; to talk of it as "nothing" simply hinders a proper scientific investigation of its properties. The ideas of space and time too have drawn many great minds into a morass of esoteric, mental hand waving. We're told about hyperspace and superspace and bits and pieces of space-time. We're assured that imaginary time eliminates the need for a beginning and that time itself is an illusion. Often mathematical models of space, which exist simply on paper, so to speak, are confused with the physical structure of space.

But before we leap off the cliffs of sanity and sink into a swamp of abstractions, let's do a reality check. We begin with space and time because quirky ideas about it lead to quirky consequences across the whole spectrum of science. If we think of time as an illusion, then we will have a radically confused idea of the world as a whole. And unlike certain space-time theories, we have to recognize the reality of agents and events, causes and effects in the world—that's if we want to do science or stay sane.

I might seem to be dismissing common experience when I say there's no such thing as space; however, I think I can show you that neither experience nor experiment furnish us with a thing called space.

1. There's No Such Thing as Space

When we hear of space, time and matter, we know what matter is. We can observe and measure it. We know it acts on other pieces of matter. Concurrently we assume that space is the medium in which matter exists, and that time relates to the continuing existence of matter in space. But if we're asked what is meant by space or time, we'll be hard put to give an answer.

Misconceptions

So here's my point about space. Space is not something that can be observed. It doesn't act on anything or cause anything to happen. It's not a medium with physical properties. It isn't something made of mass-energy. But then what is it? And what's the difference between space and empty space?

The fact of the matter is that there's no such thing as space. What we call space boils down to a way of talking about relationships between different bodies. Modern science sees bodies as fields and fields, for example the electromagnetic field, as possessing physical properties such as energy and momentum. Matter itself is seen as a field on space-time, and every physical object is seen as a space-time point. But fields are physical realities and not a collection of fictional points.

In common parlance, we talk as if space is a kind of thing or place. We say the satellite is spinning in space or the city is running out of space for new housing developments. But a little thought will show that in all these expressions space is not a something. It's a way of talking about the location of things or distances between things. A given area of space is thought of as a collection of dimensionless points. These points aren't physical things. They're simply mathematical ways of representing position. All material things, force fields and vacuums reside at particular points relative to each other. There are no actual points there, just as there are no lines of latitude and longitude criss-crossing a city. Distance is also an important fact on its own here. If a car is a hundred miles from a city we don't need to say that there are a hundred miles of space between the car and the city. We are simply saying that the car is positioned 100 miles from the city. There is a distance of a hundred miles between the car and the city. The statement that the satellite is in space can be restated to say that it's at a certain distance from the earth and is following a circular path of motion.

Space, in short, is a path along which matter can move (matter here is shorthand for energy). Dimension is the direction in which this motion is possible. There are three dimensions of space, three ways in which matter can move, that are evident in our experience: length, breadth, and height. If there is only one dimension you get a straight line, if two a plane and if three a cube. Time is spoken of as a fourth dimension, but it is certainly not a fourth dimension *of space*. There is also much speculation about additional spatial dimensions, but *currently* there is no generally accepted experimental or experiential data that supports this hypothesis.

A good analogy to space is the stock market. There's no such physical entity as the stock market. It's simply a process of exchanging transactions, of interactions between individuals. We talk of taking a company public on the stock market. In that scenario, the company isn't physically transported to the stock market. An ownership interest in the company becomes available through the network of transactions called the market. So space is the meeting-place of transactions, and is not a physical place. It's simply the transactions themselves. The other obvious comparison for space is cyberspace. Again, there's no physical location for cyberspace.

In all comments about space, we're really talking about relationships between two or more pieces of matter. The Quantum Gravity theorist Lee Smolin compares space to a sentence where matter makes up the words. The grammatical structure that defines the relationships between the words is the geometry of the universe. Obviously it makes no sense to speak of a sentence or a grammatical structure if there are no words.

Einstein's Insights

Ultimately we have a choice between saying that space is a something or that it's just another term for emptiness, i.e., nothingness. Many discussions assume that space is a kind of thing.

If it is something in this sense then it has to be something physical. Hence for a while it was held that space was a physical medium through which things moved, but which did not have an effect on things. The name for this medium was ether. But if we say that there is a physical structure of this kind, it must have physical properties and it must be detectable on a physical level. The Michelson-Morley experiment and Einstein's Theory of Special Relativity proved conclusively that the ether didn't exist.

With this theory, Einstein also eliminated Newton's idea that space was a container that existed whether or not there was matter in it. What the theory did was to show that space and time were inseparable from matter and motion. One of Einstein's great insights was his discovery that physical fields, in a sense, created space. Space, like time, comes into being with matter and motion.

Since space means simply paths along which it is possible for particles to move or fields to act, the quantity and the distribution of matter in the universe determine the shape and extent of space. We say that space is flat if light particles travel in straight lines. We say it is curved if light is bent. This geometry of space is determined by physical experiments. Our measurements tell us if space is flat or curved. And it's plainly wrong to proclaim that the curvature of space produces matter. Space can be curved because of matter and not the other way round.

Pre-Einstein, we thought of space and time as separate stages on which particles could interact; the stages would remain in place whether or not there were actors available. Post-Einstein, space-time *is* the interaction of particle-events.

Thus Einstein's theories support the position that space is simply a word we use to describe a network of relationships between fields or their manifestations, i.e. particles. Certainly space-time exists, but space-time isn't a thing. It's the relationship matrix of causes and effects, agents and events. Smolin points out that Relativity is a theory of evolving relationships with a particular emphasis on fields.

I have argued that space or space-time do not exist as substances. But even if they did exist as entities in their own right, we would still have to ask of them the same question we ask of matter: how did it come to exist and how did it come to possess the properties peculiar to it?

Finite and Infinite Space

Geek: In your view, is space finite or infinite?

Guru: You should really ask whether there is a finite or an infinite amount of matter in the universe. If there's no matter, no fields, it's meaningless to speak of space since you would only be talking about nothingness. And there's no difference between a finite and an infinite amount of nothingness. It's all nothing. You might talk of a finite or infinite range of coordinates, but then we're talking mathematics and not physical reality.

It's a different story when you talk about matter. Scientifically, well before Einstein, it was clear that a Euclidean universe couldn't have infinite matter. As Stanley Jaki pointed out, an infinite number of stars distributed homogeneously would exercise either an infinite or zero gravitational force at all points. And light from the infinite distribution of stars would shine with infinite intensity at all points in the universe; thus, there would be no night sky.

Geek: But suppose the universe is not Euclidean. Suppose its space has a geometry that can accommodate an infinite amount of matter without gravitational paradoxes.

Guru: Jaki and various mathematicians have argued that such an actual infinity is a mathematical impossibility. But the debate about whether or not it's possible for an infinite number of things to exist shouldn't detain us. There is one thing no one can deny: it's not possible for us to know that there are an infinite number of things simply because it's impossible for us to count an infinite number of things. In other words, we can't say that there's an infinite number of things, or an infinite number of universes, because if they really are infinite we can't count them to know they're infinite. Of course, there's no evidence whatsoever for the claim that there's an infinite amount of matter. Scientists are still trying to determine how much mass-energy there is in the observable universe.

Geek: You've mentioned Einstein in passing. But you haven't addressed the full impact of his relativity theories on our ideas of space and time.

Guru: Everything I've said assumes the validity of these theories. But we should consider Einstein's contributions further and I'll do that next.

Let me summarize what I've said thus far. The words "space," "time" and "space-time" have taken on lives of their own without any relation to experience or experiment. They are thought of as things on a par with mass and energy. This is not the first time that people have formed beliefs about the material world without any physical evidence. Think of the ether that was taken for granted in the 19th century or the idea so brilliantly refuted by Maxwell that heat was a kind of fluid. Only physically measurable and quantifiable things belong to science. Any discussion of things without physical properties falls outside science.

Another point of confusion is the notion of dimension, which is nothing more or less than a direction in which motion is possible. Discussions about other dimensions should not obscure the fact that (a) we cannot truly visualize four or eleven or more dimensions of space; (b) there's no experimental evidence for the existence of these dimensions, although a great desire to believe in them; and (c) the existence of additional dimensions still leaves open the question of how these additional dimensions as well as the three we know of in space came to be.

2. Matter in Motion from Newton to Einstein

Let's turn now to Einstein. I think it's no exaggeration to say that a good part of physics is about matter and motion. Hundreds and thousands of scientists have helped expand our understanding of both realms. But it was Newton and Einstein who changed the face of physics with their own giant leaps of thought. They were both visionaries driven by all or nothing theories of the universe, theories they built single-handedly on superstructures of physical laws, mathematical derivations and experimental predictions.

Newton is popularly remembered for his three laws of motion that address the relative positions, velocities and accelerations of bodies in motion. The first law relates to the constant speed of moving bodies, the second details the effect of forces acting on bodies in motion, and the third concerns the equal and opposite reactions of bodies acting on each other. His inverse square law of gravitation asserts that the gravitational force between two bodies is proportional to their masses and the distance between them. The force acts instantaneously between the two bodies no matter what the distance. Newton is also known for his view of space and time as absolute entities that exist independently of things and events. As noted, he held that space is a container and time flows without reference to anything else.

Newton's laws of motion have remained unchallenged since they were first introduced, albeit with adaptations in such exceptional situations as speeds close to the velocity of light, the behavior of subatomic particles and the gravitation of the universe as a whole. But his theories of absolute space and time, an absolute frame of rest and instant action-at-a-distance, in gravitation, were permanently displaced by the Einsteinian revolution.

Einstein introduced Special Relativity in 1905 and General Relativity in 1916. The bending of light near massive objects like the sun was one of the major predictions of General Relativity. This prediction was confirmed in the solar eclipse of 1919 and established the reputation of both the theory and its author.

For our purposes here, six of Einstein's insights are especially important. These insights concern:

a. the speed of light

b. the invariance of scientific law

c. space-time

d. motion

e. the dynamic nature of the universe

f. mass-energy equivalence

a. The speed of light

Einstein's starting point is the speed of light. The speed of light, 186,283 miles per second in a vacuum, is an absolute constant and remains the same in all directions and for observers moving at any other speed. Moreover, no energy or information can travel faster than light. As a moving body approaches the speed of light, it requires greater and greater amounts of energy to accelerate it. It would acquire infinite mass if it did reach the speed of light, but to travel at that speed, it would have to be accelerated by an infinite force. The photon, which has zero rest mass, on the other hand, can travel at the speed of light. Interestingly, Einstein's initial work on relativity grew out of his study of the behavior of light.

b. Invariance

It's sometimes said that the Theory of Relativity shows all truths to be relative. As Stanley Jaki has shown, this was hardly what Einstein set out to do, and in fact he was initially inclined to call his discovery the theory of invariance. According to the historian of science Gerald Holton, Einstein's theory essentially shows that "some things are invariant." "From the Relative to the Absolute" was the title of Max Planck's famous speech on Einstein's theories. Special Relativity says that the laws of physics are the same for all observers moving at constant speeds relative to each other. General Relativity extends this to motions that are accelerated with respect to one another and includes gravitational fields. Einstein's Equivalence Principle states that the laws of physics are the same in a uniform gravitational field as they are with respect to a reference frame going through uniform acceleration. And, as already seen, the speed of light is a fundamental constant in the Theory of Relativity. Einstein was also an absolutist with respect to the fundamental assumptions for doing science. He held that science is rooted in the belief that there is an actual world external to us.

c. Space-time

Until Einstein, it was assumed that events could be understood simply by reference to location in space. The Theories of Relativity showed that location, duration and state of motion are all inextricably tied to every event. Each event is located by four coordinates, three of space and one of time. When an event takes place depends on where it's located. For example, the images we see of distant galaxies reveal events that took place millions of years ago because that's how long it takes for their light to reach us. All events take place in a combination of three dimensions of space and one of time that's called the space-time continuum. And what we experience as duration is determined by our state of motion as, for instance, in time dilation. Events take place at points in space-time, and a history of events relating to a particle is its world-line in space-time.

The dynamic interaction of matter and space-time is called gravity. Matter curves space-time and matter travels along the curved path of space-time. If space-time is thought of as a rubber sheet that stretches out everywhere and a star or planet is seen as a bowling ball that lies on the sheet, then the curvature on the sheet is gravity. The curvature is determined by the distribution of mass-energy-momentum. Any body traveling in space is affected by its curvature and travels in the direction of curvature; this includes light, which has no mass. Einstein spoke of space-time as a structural quality of the gravitational field. It's important to remember that space-time is a mathematical idea and not something physical, although the universe itself is physical. We can't talk of traveling through space-time.

The Theory of Relativity description of space-time is consistent with the idea of space as a path of motion for a particle. It's true that space-time in Relativity may be thought off as a field, but fields can be thought of here as physical effects with specific causes. Space-time itself is the sum total of all events.

ഐ

d. Motion

All motion is relative. No moving body is in a state of absolute uniform motion because it's in motion only relative to some other body. When we talk of a particular body at rest as a reference point, this is only a matter of convention since it may not actually be at rest with respect to something else. Any object on earth, for instance, is automatically traveling through space. There is no preferred frame of reference for any moving body. And, of course, nothing in the universe is at absolute rest. A person who's apparently motionless on a couch is actually moving at a high speed with the rest of Planet Earth around the sun, and the sun itself is rotating around a galaxy and the galaxy is speeding away from other galaxies. Also, individuals who are in different states of motion experience space and time differently. These differences will become obvious at speeds close to the speed of light. Experiments have shown, for instance, that moving clocks run slowly compared to a clock at rest, and this is the phenomenon of time dilation. Bodies in motion appear to shrink. Once we understand time as the measurement of change, it should not seem surprising that people undergoing changes at different rates will also experience the passage of time differently.

e. Dynamic universe

Einstein instinctively preferred a static universe that had no beginning or end and was uniform in space. But his equation for motion in space-time required that a universe with matter could not remain static. If the universe had no matter, i.e. zero density, it would remain changeless. But once matter came into the picture, then space-time would curve, and this curvature in turn would affect the motion of matter. To avoid this conclusion, Einstein introduced a variable called the cosmological constant that could yield a static universe. But Einstein withdrew this cosmological constant, calling it his worst mistake, when confronted with actual astronomical evidence of the expansion of the universe. Assuming that matter is distributed uniformly in space, the curvature of space will be: (a) positive, like a globe without edges; (b) flat, extending without end because there's no curvature; OR (c) negative, often compared to a saddle but impossible to imagine. Universes with positive curvature expand and then contract, flat universes continue to expand and eventually stop, and negatively curved universes expand in perpetuity. The Singularity Theorem of Stephen Hawking and Roger Penrose, worked out within the framework of the Theory of Relativity, specifies that the universe had a beginning in time.

f. Mass-Energy Equivalence

With his legendary $e=mc^2$ equation, where e is the energy content of a mass m at rest and c is the speed of light, Einstein established mass and energy as interchangeable properties of matter. Mass is condensed energy, as Schroeder puts it.

Recent proposals for modifications to the Special Theory of Relativity such as DSR or doubly special relativity are still tentative at best. Its proponents say Einstein's invariant speed applies to low-energy particles of light whereas high-energy particles, they postulate, can travel faster. Although DSR is compatible with certain new cosmological findings, there is as yet no positive evidence in its favor. Hence it remains a minority position.

3. Time Out?

Our next theme is time. What is time? Is our experience of past, present and future an illusion, as some philosophers and scientists tell us? Does time have a direction, an arrow, given that all the laws of physics are equally applicable whether you go forward or backward in time? Is time another dimension of space as some scientists claim?

These questions are of obvious importance in determining our view of the world. They are not, however, scientific questions. The nature of time, I will try to show, is not ultimately determined by physics or mathematics. The sciences can talk about our experience of time under different physical conditions, for instance, the effect on it of distance or motion or gravity. But if we ask, "What is time," we are asking a meta-scientific question, one which we answer on the basis of our experience as conscious agents, not with experiments or speculative theories. So neither scientists nor philosophers set the rules in this domain. Any person of sound mind is in principle capable of giving the right answers here. The only rules are that we must make sense of the phenomena of our experience and that we must be conceptually coherent. If our concepts are confused or contradictory, we will simply end up deluding ourselves.

> *Geek*: I agree that time is a very important concept when it comes to the origin issues we've talked about. It's important because the modern understanding of time has eliminated old ideas of a

beginning of time or a singularity at the start. This is particularly clear in the work of Stephen Hawking.

Three other developments pose formidable challenges to your outlook:

1. I think the Theories of Relativity have changed EVERY-THING when it comes to time. Relativity shows there is no past, present or future since there is no preferred frame of reference.

2. Some modern physicists claim to have shown that there is no direction or arrow of time.

3. Ian Barbour has given the first scientific study of the nature of time and he comes to the conclusion that time is simply an illusion.

The whole idea that there is such a thing as time is in question, and this affects all questions about the origin or beginning of the world.

Guru: I think your succinct overview shows how important time is to our dialogue. Unfortunately, most scientific discussions in this area are marked by confusion. I agree that some scientists have reached the conclusions you speak of. But, to put it bluntly, they're simply mistaken. In fact, I hold that general relativity and quantum physics both show the reality of time.

Let me start by questioning the general assumption that science is the right tool for studying time. You might be interested to know it was Einstein who once remarked to Rudolf Carnap that the reality of the now is beyond the competence of physics. It's remarkable that Einstein, the scientist who has had more influence on our understanding of time than anyone else, held that time cannot be treated scientifically. We need to keep this in mind as we proceed.

Obviously time is not a physical quantity that can be observed with appropriate devices or reduced solely to mathematical formulae! Devices can measure time with increasing sophistication, via the oscillation of atoms for instance, but they can't tell us the meaning of time. Science, of course, furnishes us with a wealth of data that can be applied in answering questions about time. But the answers themselves are interpretations of the data. The interpretations are meta-scientific because they rely on dimensions of our experience that go beyond scientific experimentation.

My approach will be to flesh out what I think we can know about time, see how this relates to science, and finally consider the work of Barbour and Hawking. Curiously, a lot of the work on time seems to be coming from Britain. Both Barbour and Hawking are British. The two thinkers who, in my view, have brought the most clarity to discussions of time are also British, John Lucas and David Braine.

What is Time?

At a physical level, time is a measurement of change. The upshot of the theory of relativity is that space and time are inseparably tied to matter and motion. But this is nothing new. As Hawking points out, St. Augustine had seen the connection centuries ago when he said that the universe is created with time and not in time. In the context of relativity, Lee Smolin notes that time is to be seen as a measure of changes in the network of relationships we call space, and neither time nor space mean anything outside these relationships.

Science revolves around providing explanations for events, and the scientific method is based on cause and effect relationships. Every cause and effect relationship involves an agent that causes an event. The event is the effect, e.g., if fire and gunpowder come together, then the result is an explosion. An agent, as defined by David Braine, is someone or something with active power to cause events and other effects in the field in which it acts. Agents can be persons, life-forms, fields, particles; they can be intentional or inanimate. Now the relationship of cause and effect linking agents and events is not possible if there's no before and after. When we talk of some agent in physics causing an event to take place, we explicitly assume there's an interval of time, no matter how short, between cause and effect. Time here is real and has a definite direction. Time springs from agents, events, and relationships of cause and effect brought about by agents. There's no time independent of agents and events.

The Arrow of Time

Historically the so-called direction or arrow of time has been called into question in science. This issue puzzled Einstein and scientists before and after him because scientific equations have no "before" and "after" variables on each side of the equation. In classical and quantum mechanics and in relativity, time is symmetrical and has no direction. In thermodynamics, there is an apparent direction with the increase of entropy, the irreversible increase of disorder, in a closed system. But we don't know what level of entropy existed at the origin of the universe, and so the argument that entropy shows an arrow of time is at least disputable.

If you play a movie in reverse, it will be obvious that you're seeing things in a way that couldn't be the case. There's an obvious direction to time. But if you reduce all the same activities to their quantum level, then everything comes down to the motion of subatomic particles. And from viewing this motion, we cannot detect any differences between past and future. The motion can proceed in either direction without violating our sense of what's possible. So does scientific law show there is no irreversible arrow of time?

In addressing this issue, Braine admits that scientists may think of science as simply a collection of laws and equations that do not involve time—past, present or future—as a variable. But every law is derived from the observation of events and agents. These events range from the history of life on earth to experiments with devices carried out in a specialized setting. Every such event involves a cause and an effect brought about by agents and a before and an after. Every cause comes before the effect, and so the time element is essential. The laws themselves are predictions that certain causes will bring about certain effects, and this involves time through and through. Even quantum physics is about agents (waves and particles) exercising power in a cause and effect relationship, although we can't visually imagine these agents.

John Lucas even argues that quantum physics offers the most fundamental evidence for the reality of time because of its inherent directedness. In the quantum realm, as we shall see, there is a clear difference between before and after in quantum events, such as the collapse of the wavefunction. The present is what actually happens and the past is thereafter unalterable. Several possible scenarios coalesce at a certain point into one state of affairs—and there's no going back.

Some scientists have interpreted the theory of relativity to imply that our experience of time is an illusion because what's present in one frame of reference is future in another, and so there's no absolute difference between past, present or future. Such divisions are purely subjective. But this train of thought is mistaken because relativity tells us that some events lie in the past and others in the future no matter what frame of reference we use. It's true that the Special Theory of Relativity says there's no absolute simultaneity throughout the universe. But it's certainly compatible with this theory to hold that there's an absolute before and after between cause-and-effect relationships. Braine notes, in fact, that the Special Theory of Relativity, with its principle that no influence can travel faster than light, reinforces the idea that an effect has to come *after* the cause.

Relativity, we have seen, shows us that time is not a container that flows from past to future, independent of agents and events. Relativity reinforces the idea of time as a web of cause and effect relationships involving agents and events. *Agents are real, events are real, causes and effects are real, and so time is real.*

And there's a lot more to time than just the laws of physics. Braine points out that biology and history show a direction in which order is created. For instance, when a composer decides to author a symphony, there is a sequence and a direction from the initial intention to the final production.

Lucas shows that any theory of time has to cope with certain features of experience that are undeniable. The best explanation of our experience of time, chiefly change, duration, the fact that events don't all take place simultaneously, the order of past, present and future, is that things really are this way. And without real linear time, we can't make sense of our continuing personal identity. To be conscious of things happening in a sequence and to act as an agent, say, in communicating with another, is to assume that time is real. We couldn't be scientists, says Lucas, unless we're also agents, agents who observe and experiment and exercise choice.

Barbour's End of Time

Geek: I guess a lot of this is a matter of interpretation. But the two best-known theories of time of our day, developed by two great scientists, reach fundamentally different conclusions from yours using the same data. They claim to have demolished the basis of traditional beliefs. I'd like to know how you respond to them or at least their arguments.

Guru: Yes, it's a matter of interpretation. But the only interpretation that deserves to be taken seriously is one that both makes sense of the data of experience and provides supportive evidence. Despite the eminence of its physicist-authors, the theories of Barbour and Hawking fly in the face of our immediate experience and, just as important, there's not the slightest shred of evidence in their favor. In their ruminations on time, Hawking and Barbour are speaking more as philosophers than as scientists, and I will address them at this level.

In *The End of Time* Barbour presents the extraordinary, and unbelievable, view that time doesn't exist. Our experience of change or motion is all an illusion, as is our memory of the past. The only things that exist are instants of time or "Nows" called "time capsules" that are instantaneous arrangements of all the things in the universe. These capsules are static and timeless containers of records, but they're experienced as dynamic and temporal and give the illusion of change. We're locked in the present, and that's the only thing that's real. Our belief that there's a succession of events is an illusion. There is no such thing as history.

I started off saying that Barbour's view was unbelievable. I hope you see why. In my view it's not a position that can be taken seriously by anyone who takes either science or everyday living seriously (it's a different matter for those who seek novelty for its own sake). On a fundamental level, Barbour has three insurmountable problems.

First, by denying change and motion, he's denying something as basic as the existence of the world around us. Certainly, we can claim that the world doesn't exist, but any argument that challenges the existence of the world already assumes that it exists, since the argument has to be made with things in the world like words in some physical format. Likewise, it has been pointed out that, in arguing there's no change, you assume that you can change someone's mind or brain-state. Your argument must change the *same* brain-state that believes in change to a brain-state that doesn't believe in change. But this contradicts the argument that there's no change.

Secondly, Barbour has to explain away something that appears obvious to us, namely our experience of motion and change. But his attempted explanation is wholly implausible since it relies on the purely speculative idea that instants of time have a special structure that misleads us into believing in change. The problem is not simply that it's a piece of speculation that denies everyday experience, but also that no evidence is offered to support it.

Thirdly, not only is there no experimental evidence in support of the theory but Barbour admits he cannot think of any direct experimental way to test its truth. A prominent theorist of time, Jeremy Butterfield, has carried out a thorough analysis of Barbour's proposal. He points out that there's no support for the denial of time in classical physics or in the four main approaches to interpreting quantum physics. Besides, the whole theory depends on Barbour's conjectures in Quantum Gravity (the attempted unification of quantum theory and Relativity), conjectures that lie entirely outside mainstream thought in this area. Barbour admits he has no hard and fast arguments, let alone proofs, in this domain that is so foundational for his theory. Ultimately the theory is a metaphysical and not a scientific proposal. And it does not work as metaphysics because it cannot plausibly be tied to experience.

> *Geek*: I have to admit that Barbour's thesis sounded fairly far-fetched on the face of it. But the same isn't true about Hawking, although many of those who cite his conclusions have no idea how he reached them. I'd say your fiercest challenge comes from Hawking, with his conclusive critique of belief in a supernatural origin of the universe.

4. Hawking's Time-Less History

Guru: Without question, Hawking is one of the premier contemporary expositors of modern science. It's true that, in his best-selling *A Brief History of Time,* he follows up his own proposal for the origin of the universe with the question, "What place then for a creator?" But Hawking is by no means hostile to religion. He has remarked more than once that he has left open the issue of God's existence. He ends his *Brief History* with the comment that his or any mathematical model can describe the universe but cannot answer the question of why it bothers to exist. In a recent interview with *Reason,* he speculates that God could be the ultimate determinant of which equations should materialize in a universe. So he's no enemy of a religious view of the world. But in view of his influence on our domain of discussion, I will address his central challenge.

Standard theories of the origin of the universe assume that current formulations of the laws of physics break down at the Big Bang, which is called a singularity because it's a singular event. The mass/energy of the universe will be infinitely dense at that point. In fact, as Hawking notes in *The Universe in a Nutshell*, it was he and Roger Penrose who proved that, in the mathematical model of General Relativity, time should have a beginning called the Big Bang.

But physicists are not comfortable with singular events and Hawking, among others, has tried to eliminate the idea of an initial singularity. By deploying a mathematical model used in quantum physics, Hawking claims to show that the universe has no boundary, edge, beginning or singularity.

His approach meshes with an ongoing effort among physicists to combine quantum and classical physics in describing the origin of the universe. A unified theory of this sort, called Quantum Gravity, is required because the universe at its inception was a quantum particle of incredible density. The transition of the universe from a quantum to a classical state presently belongs to the realm of speculation, and Hawking's model is one of several proposals.

His model suggests that there's no earliest moment of time because time is one more dimension of space prior to the so-called origin of the universe, and it makes no sense to ask time-type questions about a dimension of space. Thus the Big Bang can no longer be considered a singularity at the dawn of time. There is no dawn of time. The history of the universe is akin to the surface of a ball. Just as the ball has no points or edges, the universe has no first point of time, no edge to space-time, no boundary. To ask what happened before the Big Bang is like asking what is north of the North Pole. Hawking's conclusion is based on key assumptions:

a. He takes the idea of the wavefunction of a particle from quantum mechanics and applies it to the whole universe. The wavefunction of a quantum particle is information on its position and momentum, based on probabilities. When it comes to the wavefunction of the universe, at the simplest level its size may be thought of as its position and its rate of expansion its momentum. At the quantum level, the history and future state of a particle is determined by analyzing all the possible paths in space-time it could take.

b. To perform these calculations, quantum physicists have introduced the concept of imaginary time. For the purposes of their equations, time is treated as another dimension of space with the understanding that this is purely a mathematical device without any physical counterpart. They didn't think time became space; once the calculation was completed, time would return to its normal standing. In order to derive a wavefunction for the universe that eliminates singularities, Hawking applies this procedure of turning real time into imaginary time. By multiplying the time coordinate with the square root of a negative number, minus one, it becomes a coordinate of space, a direction in space. The distinction between time and space disappears and so does the idea of a beginning of time. This four-dimensional space is compared to a two-dimensional globe that has no boundaries or beginnings. A sphere has no singular points. Although there is a beginning of real time in the Big Bang, the laws governing this beginning are determined in imaginary time, and so no appeal has to be made to any external agency for the event. A universe of finite size emerges from a quantum fuzz.

d. Hawking concludes that the elimination of singularities and initial conditions results in a self-contained universe without beginning or end, and no role for a Creator. Crucially, he grants that God could have a role in determining the laws of the universe and that, even with a unified theory, there is still the question of how rules and equations are converted into a universe.

There are four levels at which Hawking's proposal can be considered:

1. First, and most fundamental, its coherence

2. Second, the evidence in favor of it

3. Third, the validity of its mathematical framework

4. Fourth, the physical embodiment of the theory.

Our focus will be on the first two levels, although there is much controversy over the last two as well.

The central problem with Hawking's account of the origin of things is that it doesn't pass the coherence test. This may seem to be a somewhat odd judgment but it's directed at his metaphysics and not his physics. And I think his errors here should seem obvious to anyone who goes behind the mathematical and quantum mechanical backdrop to his meta-scientific claims. As you will see, my responses below center on the meta-scientific assumptions and arguments and not the scientific theorizing as such.

1. Hawking claims that at a certain point time was actually another dimension of space, and hence the question of the beginning of time does not arise since you don't ask questions concerning beginning about a dimension of space. Instead of three dimensions of space and one dimension of time, we end up with four dimensions of space. These four dimensions are compared to the earth, which has no boundary or edge. Now it's one thing to say that the metaphor of time as space is applicable in making a calculation, quite another to say that it describes the physical world. And the latter seems to be what Hawking is saying; he suggests, for instance, that imaginary time may be more basic than real time. But no matter how the trick is pulled off, we can be sure it is a trick. When asked in *Reason* whether his imaginary time is a sort of mathematical trick, Hawking naughtily replies, "Yes, but perhaps an insightful trick." We know for a fact that we experience time as something distinctly different from a dimension in space. To move up or down, to the right or the left is to talk of space. To talk of today and tomorrow, before and after, is to talk of time, and this experience of things taking place in sequence is not an illusion. Any claim of actually turning time into space is as far-fetched as the idea that a law of motion could suddenly turn into a particle.

2. Two of Hawking's most penetrating critics, Keith Ward and William Lane Craig, have shown other problems with the model.

   Ward notes that Hawking replaces the human experience of time with a mathematical variable that in turn is changed into an abstract concept; the end-result cannot be called time in any real sense. Moreover, all this arises not from nothing but a world of abstract mathematical entities that is even more mysterious than the universe itself. The mathematical world cannot cause the physical, nor can it exist before the physical, since time is supposed to have been eliminated.

   Craig contends that the idea of imaginary time is physically unintelligible; an imaginary period of time makes as much sense as an imaginary number of people in a room. Hawking cannot therefore eliminate the singularity of the Big Bang since his vehicle for doing so, imaginary time, is itself unintelligible in the physical world. Like Ward, Craig points out that it's incoherent to speak of the world existing in a timeless era before time. If this era existed before time then it is in time. To save the theory, you have to step out of real time into imaginary time and then back and forth at various times. The atheist Quentin Smith has in fact called Hawking's application of imaginary time preposterous because the universe very obviously lapses in real and not imaginary time.

3. On the question of evidence, Hawking admits that his theory is at best a proposal motivated by aesthetic and metaphysical preferences. There is no hard evidence supporting it, although, like other speculative proposals, it is, arguably, not incompatible with the data. He also starts with the assumption, unsupported by data, of a no-boundary universe; this is an assumption he has to make to get the answers he wants. The process is circular. One observer points out that quantum cosmology in general tries to support mere aesthetic preferences by generating technically complex constructions with very little physical justification.

4. Hawking's collaborator George Ellis has said that Hawking's proposal assumes that certain riddles in quantum theory have been resolved along the lines he proposes. But issues like the collapse of the wavefunction, so crucial to Hawking's speculation, are still unresolved. When the role of the observer and the application of probability in measuring the wavefunction of an electron are still in debate, it seems quite a stretch to make claims about the wavefunction of the universe.

5. Another commentator, Robin Poidevin, who is, in fact, an atheist, suggests that Hawking's model of a closed universe makes the implicit, and bizarre, claim that all causes are causes of themselves. Hawking holds that the universe existed for a finite time. This universe is a closed loop of causes and effects with no first cause responsible for either any other cause or the series of causes as a whole. But under this scenario every cause is a cause of itself—an obvious impossibility. It could be argued that the

only way to explain the existence of the finite series of causes is to point to a cause lying outside the series. The philosopher Richard Swinburne has said that Hawking's interpretation of time involves the assumption that time is closed and therefore cyclical; thus, if you live long enough after 1995 you might come from 1994 into 1995 again. This cyclical view of time he finds to be contradictory (tomorrow is both after and before today) and therefore false.

If nothing else, Hawking's treatment of time illustrates the importance of our view of time. In both *A Brief History of Time* and *The Universe in a Nutshell* Hawking describes himself as a positivist. He seems unaware of the fact that Positivism, the idea that only what is scientifically verifiable is real, is among the most discredited philosophies of recent times. I hope to presently show what's wrong with Positivism. But what's relevant here is this: Hawking frankly admits that, being a Positivist, he can only give mathematical models of time and predictions associated with these models, but he cannot say what time actually is! With this starting point, it's hardly surprising that he goes off the rails when he talks about the nature of time.

Having said all this, I want to reiterate my earlier point that Hawking certainly doesn't deny the possibility of an infinite Intelligence who turns mathematical models into a universe. In *Black Holes and Baby Universes*, he notes that his new theory of time assumes "that the way the universe began would be determined by the laws of physics. One wouldn't have to say that God chose to set the universe going in some arbitrary way that we couldn't understand. It says nothing about whether or not God exists—just that He is not arbitrary." There would be plenty for a Creator to do even if the universe were fashioned on the lines laid out by Hawking. The universe did not emerge from sheer nothingness in this model. There are laws of nature that govern quantum fields and the systems that transform a quantum into a classical universe. And if we ask about the origin of these laws, the question of the Mind of God, introduced by Hawking at the end of *A Brief History,* is back on the table.

❧

Chapter

The Quantum Dimension

*T*HE CONSIDERATION OF THE UNIVERSE AS A WHOLE is the first stage of our journey to infinite Intelligence. The next step is an inquiry into the microverse, the quantum world. For clarity's sake, I will provide first a historical sketch and then an interpretation of the data.

The first three decades of the twentieth century marked the golden age of modern physics. These were the years when a sudden but sustained spurt of discoveries changed our whole understanding of physical reality. These breakthroughs ranged from a revolutionary understanding of the very large, relativity theory, to a startling vision of the very small, quantum physics. By the end of the nineteenth century some physicists had already begun speaking of the end of physics, noting that only two puzzles remained to be solved: (a) the Michelson-Morley experiment that failed to show the existence of the ether and (b) the phenomenon of black-body radiation. Curiously enough, it was precisely these two problems that gave birth to the theories of relativity and quantum mechanics, and rewrote the roadmap of physics in the next century.

છ્ય

1. The Quantum Leap

Quantum physics is for some the holy of holies. They speak of it in hushed tones with frequent genuflections before a tabernacle of fundamental uncertainties, collapsed wavefunctions and brave new ideas such as non-locality and entanglement. Appeal to the authority of the Quantum silences all argument and overpowers the temptation to turn to rational thought, and the much-disdained two-value logic.

My comments above are by no means a parody of the serious and fruitful research carried out by multitudes of scientists. I'm simply drawing attention to the damage caused by enthusiasts who wear the quantum cloak to solemnly proclaim fanciful notions. We are told that every particle creates an infinite number of parallel worlds. It is claimed that something exists only if it is observed. Along with the scientifically verifiable findings of quantum theory, these whimsical ideas too are hailed as quantum breakthroughs.

In quantum physics, as in several other key flashpoints of science—such as space and time, the Big Bang, the origin of life—philosophy, rather than hard science, is in the foreground. Scientists are making philosophical arguments using the language of science, drawing philosophical conclusions unrelated to their scientific data and eagerly breathing life into skeletons in the philosophical closet. Four major meta-scientific issues are at stake in this discussion:

- ♦ Does the world exist or is it simply a figment of our consciousness?

- ♦ Can we know anything?

- ♦ Is there a principle of cause and effect or is there no such correlation?

- ♦ Can something come from nothing?

Such popular presentations of quantum theory as the famed Copenhagen Interpretation and the Many Worlds hypothesis, both of which will be described below, are primarily philosophical interpretations of subatomic phenomena. At times the practitioners know they are giving their own views on the data and at others they fail to make any distinction between fact and speculation. But one primary rule is observed more in its violation than its enforcement: scientific data cannot in and of itself serve as a philosophical argument or proof. In other words, none of the meta-scientific issues mentioned here can be resolved by simply drawing attention to some quantum phenomenon or formula. The fact that a physical interaction can't be measured exactly doesn't warrant the conclusion that such an interaction didn't actually take place. It may very well be the case that such a measurement isn't physically possible and so there's no point insisting on the existence of a hidden variable that will some day enable measurement. But the fact that something can't be exactly measured doesn't mean that it doesn't exist. Unfortunately, neither the proponents of fashionably outlandish interpretations nor their scientific critics really seem to have understood this point.

> *Geek*: I'm sure you don't disagree that the Copenhagen version of quantum physics has been one of the most successful theories ever proposed. Of course, it leaves you without an objective world of cause and effect and thus undermines the idea of a First Cause.

Guru: Copenhagen is certainly the most influential interpretation. But it's not necessarily the last word on all quantum questions. You cannot do science, quantum or otherwise, if you adopt all the Copenhagen assumptions. The distinguished philosopher of science Karl Popper held that the Copenhagen criticism of the cause-effect connection had a crippling effect on research, because the search for these connections was dismissed as meaningless. There is, in fact, no one Copenhagen position; Bohr, Heisenberg and John von Neumann all had radically different interpretations. And even the Copenhagen pioneers admitted the distinction between science and philosophy in their approach.

I'm not advocating any particular school of quantum interpretation. I'm simply suggesting that the starting-point of science has to be the existence of a real world that we can both know and explain in scientific terms. And every issue and application should be tested on its own merits. Since context is so critical, I will consider the history of quantum physics and its basic concepts and then discuss specific issues.

ᏇᎧ

2. Just the Facts

Our historical review may help us distinguish what's scientific from what's philosophical and extrapolations from hard facts.

Quantum physics began in the year 1900 with Max Planck's proposal of the quantum of action. The major contributors to quantum theory were Planck, with his discovery that energy is emitted in discrete units called quanta; Albert Einstein, with his application of Planck's theory to explain the photoelectric effect; Niels Bohr, who applied quantum ideas to the structure of the atom and founded the Copenhagen School of Complementarity, which holds that seemingly contradictory phenomena at the quantum level are actually complementary; Werner Heisenberg, best known for both his Uncertainty Principle, the notion that an observer cannot know with certainty both the position and the momentum of a subatomic particle since any measurement will require light to hit and thereby disturb the particle, and matrix mechanics, a system he developed to describe the behavior of quantum particles; Erwin Schrödinger, inventor of wave mechanics, another structure for describing quantum behavior that was mathematically equivalent to Heisenberg's system; and Louis de Broglie, Wolfgang Pauli and Paul Dirac.

In their own distinct ways, each one of these scientists played colorful roles in the rise of quantum theory. If Max Planck was the Founding Father, Bohr was the heir apparent, Einstein the estranged godfather, Heisenberg the *enfant terrible* and Schrödinger the eccentric uncle.

The theory had its roots in the attempt to explain certain anomalies in the radiation of energy, in particular the phenomenon of black body radiation. A black body is one that absorbs all radiation while radiating energy at frequencies that rise with temperature. Physicists were puzzled to find that the wave theory of electromagnetic radiation could not explain the ratio of wavelength to intensity of black body radiation. On the evening of October 7, 1900, Planck solved this problem and gave birth to quantum physics by hypothesizing that energy is radiated in chunks or bundles called quanta. When a molecule enters a lower energy state, it emits a quantum of energy. This explains variations in wavelength. Planck drew a correlation between the energy of a quantum and the frequency of radiation whereby energy=frequency x *h*, a small unit that acted as a constant. The constant, given as *h*, was later called Planck's constant, and it was a key constant that was to play the same role that *c*, the speed of light, played in relativity.

Planck's hypothesis wasn't initially accepted because the prevalent view held that nature was a continuum, not a collection of discrete quantities. In a paper in 1905, Einstein substantiated Planck's hypothesis when he used it to explain another puzzling phenomenon involving light, the photoelectric effect. It was well known that metals lose electrons when exposed to high-frequency, e.g., ultra-violet, light, a phenomenon that did not fit in with the wave theory of light. Einstein explained this effect by applying Planck's theory to light in this way: light travels not as waves but as electromagnetic quanta called photons. When high-energy photons, found at higher frequencies, strike electrons on the metal, they eject the electrons. In 1921 Einstein won the Nobel Prize for this discovery. In 1923 Arthur Compton and Peter Debye, working independently, experimentally established the existence of the photon as a particle with energy and directed momentum. Nevertheless, such phenomena as diffraction and interference required a wave model of light, and this gave rise to the idea of wave-particle duality whereby light sometimes acts as a wave and at others as a photon.

Energy continued to play a role in the genesis of quantum theory with Bohr's quantum model of the atom in 1913. Bohr moved quantum theory beyond electromagnetic radiation to the atomic structure that underlies all matter. In 1911, Rutherford had presented a solar system-like model of the atom with electrons orbiting a positively charged nucleus. However, under the canons of classical physics, specifically Maxwell's equations that describe the production of electromagnetic energy by accelerating charged particles, the orbiting electron should eventually lose energy and plunge into the nucleus. Bohr, who briefly joined Rutherford in Manchester, proposed that this does not happen because the orbits of the electrons are quantized. In a series of papers in 1913, Bohr proposed that electrons circle the nucleus with certain fixed energy levels or states. These electrons neither emit nor absorb energy except when they leap from one orbit to another. The energy radiated in a quantum leap is equivalent to the difference in energy levels between the two orbits. An electron at the lowest energy level or orbit is at its ground state. Among other things, Bohr's theory explained the frequencies of light emitted by hydrogen by relating it to the energy differences between orbits.

Louis de Broglie in 1924 advanced the idea that wave-particle duality was true not just of light but also of other particles like electrons and protons. The validity of this hypothesis was demonstrated in electron diffraction experiments in 1925 where electrons were seen to behave as waves. At the most basic level, then, matter was to be thought of as both a particle and a wave with electrons and other subatomic particles acting like stationary waves that occur with discrete frequencies.

Louis de Broglie's wave-particle idea was further established on a theoretical level in early 1926 with Erwin Schrödinger's famous wavefunction equation and the domain of wave mechanics that he created. Schrödinger believed that subatomic entities were not particles but waves in an electromagnetic field that sometimes showed particle properties. Electrons and other matter waves had crests, troughs and interference patterns and could be represented by a mathematical equation that controls each wave's propagation, its wavefunction. Schrödinger, of course, believed that the equation represented actual waves.

Schrödinger's approach was criticized by Max Born, who accepted the idea of a wavefunction but held that it refers to a probability wave—the probability of finding a particle in a given region. It is not something real in itself but pertains to our knowledge of a system. A subatomic particle is a real particle with measurable properties, such as location, momentum, energy, and its wavefunction is a mathematical formula that spells out the probability that the particle is at a particular point. The wavefunction of an electron orbiting a protonic nucleus is defined in terms of the probabilities of it being present at various spatial coordinates. These coordinates are mapped relative to the origin of the coordinate system, which is the proton at the center of the atom. Once the particle is located through measurement, the wavefunction is said to collapse because the probabilities have now become an actuality. To put it another way, once a particle's position is known through measurement its wavefunction has collapsed.

In the same year, 1926, Werner Heisenberg, a rival of Schrödinger, published his own mathematical description of the quantum world that is now called matrix mechanics. In this system, he restricted himself to physical quantities that could be observed (masses, charges) and represented these in matrices, numbers arranged in different formats (rows, columns) from which you could generate other measurable quantities (e.g., frequencies). While Schrödinger claimed his wave mechanics described the underlying reality in quantum physics, Heisenberg held that his matrix mechanics was simply an algorithm to correlate the results of experimental data and generate new predictions. Meanwhile, Paul Dirac developed his q-number algebra version of quantum mechanics that again handled such observables as position, momentum and energy. Amazingly, all three theories were mathematically equivalent, different forms of the same conceptual constructs. Dirac also developed a transformation theory with rules for turning one scheme into another.

By the end of 1926, Schrödinger's formulation of quantum mechanics was gaining ground over alternative interpretations. But Heisenberg and Bohr were determined to bring their emphasis on the non-classical nature of quantum mechanics to the foreground. After consulting with Bohr, Heisenberg in 1927 introduced his famous Uncertainty Principle. Initially his theory specified that it was impossible to simultaneously measure the position and momentum of a subatomic particle because two kinds of mutually exclusive instruments would be required. Under pressure from Bohr, he amended this formulation to make the claim that the wave-particle nature of these entities made it impossible to apply classical concepts such as wave and particle at the quantum level; that is, it is not just the case that we cannot measure their properties, but we cannot even in principle know them. For instance, if light is seen as a wave, there's uncertainty about the position of a particle under observation since there's a limit to the resolving power of light; if, on the other hand, light is viewed as a stream of photons, there can be no certainty about the momentum of the particle, since the photons give momentum to the particles they strike, and we cannot know how much of this momentum comes from the photon. A long-wavelength, low-energy photon doesn't significantly affect the momentum of the particle, but it can't give an accurate reading of its position. A short-wavelength, high-energy photon, on the other hand, gives the position accurately but not the original momentum of the particle. Heisenberg's Uncertainty Principle, with its equation for quantifying indeterminacy, coupled with Bohr's Theory of Complementarity, established what was called the Copenhagen Interpretation as the dominant view in quantum mechanics.

Other breakthroughs in the early years of quantum theory included Wolfgang Pauli's Exclusion Principle, which declared that electrons cannot share the same configuration, i.e., equal values for their quantum numbers (e.g., place, orbit), in an atom, and Dirac's equation for the angular momentum or spin of electrons, their apparent self-motion. In his 1928 paper on the relativistic quantum theory of the electron, which included the spin equation, Dirac predicted the existence of antimatter, antiparticles that had the same mass but an opposite charge of all particles. This theory was confirmed within four years with the discovery of the first antiparticle, the positron. The upshot of the theory is that every kind of particle has an antiparticle that corresponds to it. Particles and antiparticles have the same mass and similar properties but opposite charges. Matter and antimatter particles annihilate each other, but in the early history of the universe matter came to pre-dominate; today, antimatter is found not just in the laboratory but also in cosmic rays.

3. Foundations and Applications

When the smoke cleared from three decades of continual discovery, certain ideas emerged as the pillars undergirding quantum theory. The theory itself was the mathematical framework generated to describe the subatomic world. Moreover, practical applications of the theory appeared almost immediately.

Dr. Jekyll and Mr. Hyde

Subatomic particles show properties of both waves and particles. For instance, in experiments, photons sometimes act as particles and at other times as waves. The Copenhagen School tells us that a quantum entity is intrinsically indefinable until measured. The idea that it's a particle or a wave is simply a classical habit of thought that has no relevance in this realm. The entity takes a definite form only when it's measured; it has no intrinsic properties and behaves differently in different circumstances. Prior to measurement it exists as a "superposition" of all the states permitted by its wavefunction. The superposition principle is based on the idea of adding the amplitude of each wave. A measurement that differentiates between these states causes a collapse of the wavefunction that forces the entity into one final state.

Particle physicists, of course, continue to refer to the objects of their studies as particles. But these are not classical particles. Heisenberg envisioned a scenario where particles at the quantum level would be mathematically represented with descriptors like mass and spin. This is the approach taken by particle physicists who create elaborate classification schemes to describe both these entities and their interactions. The famous Schrödinger's Cat paradox explores the consequences of these concepts.

O Electron, O Photon, Where Art Thou?

A precise description of the state of a quantum system requires definite values for such variables as location, momentum and energy. But these values cannot be known with exactitude because of the limits on our ability to measure them. The Copenhagen school would go even further and say that definite values are not possible because the state of every quantum system is objectively indefinite in and of itself. Probabilities for these values can, however, be defined.

Heisenberg's Uncertainty Principle was quantified in a formula: the uncertainty of what can be known of the momentum of a particle multiplied by the uncertainty of its position is greater than or equal to Planck's constant divided by 4 times *pi*. Or as Peter Hodgson puts it, the product of the uncertainties in position and momentum is always greater than Planck's constant divided by 4*pi*.

Non-locality and Entanglement

Non-locality and entanglement are puzzling quantum phenomena that a distressed Einstein called "spooky action at a distance."

Locality declares that no signal can be transmitted faster than the speed of light; this is a consequence of the Special Theory of Relativity. Non-locality refers to the possibility that correlations could take place between spatially separated events or entities at speeds faster than light.

Entanglement is the idea that the components of a quantum mechanical system when split up continue to exercise an influence on each other, i.e. non-locality, without regard to distance. The experimental evidence of such an influence is found only in events or entities that both originate in the same quantum system. For instance, experiments have shown that an alteration in the properties, e.g., polarization/orientation or momentum, of one of a pair of photons shot from a single source causes a corresponding change in the property of the other, no matter how far away it is. If the correspondence in changes took place when the particles were together, this would not be remarkable; for instance, conservation of momentum would dictate that if one heads east, the other would go west. But experiments have shown that the behavior of one particle continues to influence what the other does even after they have been separated.

In 1964, John Stewart Bell put forward a framework for experiments later called Bell's Theorem to determine whether or not classical or quantum descriptions are more accurate with respect to quantum phenomena. Alain Aspect, in 1982, conducted a series of experiments within this framework to establish non-locality as a real phenomenon.

Scientists have argued that non-locality does not, in fact, violate relativity because it is not possible to exploit non-locality to enable faster than light communication. Non-locality does not permit transmission of information.

Mathematical Formalisms and Statistical Laws
Almost from the beginning, quantum theory has been a mathematical tour de force. The three major formulations of quantum theory, those of Schrödinger, Heisenberg, and Dirac, were mathematical masterpieces. Most concepts, events and laws in quantum theory are built around mathematical abstractions and formalisms. Because of this, and also because quantum phenomena are not directly experienced, there is very little, if anything, for the visual imagination to work on. Nevertheless, a vast body of experimental data shows the validity of these thought-creations.

Another key element is the role of probability and the consequent dependence on statistical laws. Such laws can predict the frequency of occurrence of certain events, but they cannot specify with certainty if or when an individual event is going to take place. The one-to-one correlation of laws to individual events characteristic of classical physics is simply not to be found, or expected, at the quantum level. But, again, almost all the available evidence indicates that these statistical laws work.

The Standard Model
The findings of quantum theory have given rise to a new way of viewing

the physical world. The excitations of the quantum field manifest themselves as:

♦ elementary particles

♦ the four fundamental forces or interactions

♦ the messenger particles that mediate interactions.

This infrastructure of interactions and entities, as we currently understand it, is described by what is called the Standard Model.

The quarks and leptons that constitute matter appear in a total of 12 known variants. There are six kinds of lepton: the electron, the muon and the tau with neutrinos corresponding to each of these; and six kinds of quark: up and down, strange and charmed, top and bottom. Each of the quarks is in one of three internal states: red, green and blue. These descriptions refer to their charge and not their color. Quarks are invariably housed within larger particles like protons, neutrons and mesons. In addition to quarks and leptons, there are the bosons: the W, Z and Higgs bosons and two massless bosons, the photon and the gluon.

In the everyday realm, quarks and leptons form atoms, and atoms in turn form molecules. An atom is the fundamental unit of a chemical element. At the center of the atom is the nucleus, which is made up of protons and neutrons; protons have a positive charge whereas neutrons have no charge. A cloud of electrons with a negative charge surrounds the nucleus. 99.9% of the volume of the millionth of a millimeter-size atom is an empty vacuum, with most of the mass of the atom coming from protons and neutrons. The mass of the protons and neutrons is about two thousand times greater than that of the electron.

There are 92 different kinds of atoms. These are classified in the periodic table of naturally occurring elements under the categories of atomic structure and physical and chemical properties. The simplest natural element is hydrogen, which has one proton. The most complex is uranium, which has 92 protons. Atoms generally, but not always, have the same number of neutrons as they do protons. Atoms combine with each other in specific configurations to form molecules. For instance, sodium combines with chlorine to form salt, and two hydrogen atoms combine with one oxygen atom to form water.

As shown below, the fundamental particles are subject to four fundamental forces—gravity, electromagnetism and the weak and strong nuclear forces—that both transmit energy and bind matter. These forces are mediated by messenger or exchange particles so that the interaction of particles is itself an exchange of particles: two electrons repel each other by exchanging photons.

+ **Gravity**—An attractive force between any two objects with mass. It is the weakest of the four. But despite its negligible influence at the particle level, it plays a dominant role at larger scales because of the cumulative force of particle aggregates. It is thought to be mediated by the graviton, a particle that has yet to be identified. The strength of the gravitational force between two interacting bodies is proportional to the product of their masses and the inverse square of the distance between them.

+ **Electromagnetism**—Electricity and magnetism are manifestations of a single force, electromagnetism. Like charges repel and unlike charges attract, and this force of attraction and repulsion is electricity. The flow of current or charge creates a magnetic field. The electromagnetic force is the second most powerful of the forces and is mediated by photons. This is the force that makes chemistry possible because it keeps the electron tied to the nucleus. The strength of the electromagnetic force is also a function of the inverse square of distance, but it is determined by the charges of the bodies, not their masses.

+ **Strong nuclear force**—A force that holds protons and neutrons together in a nucleus and binds quarks in protons and neutrons. The strongest of the forces, it is mediated by gluons.

+ **Weak nuclear force**—A force responsible for the radioactive decay of nuclei that is mediated by the W and Z bosons. In beta minus radioactivity, a neutron decays into a proton while emitting an electron and an electron-antineutrino. In beta plus radioactivity, a proton is transformed into a neutron with the emission of a positron and an electron neutrino. Although it is the second weakest force, and is millions of times smaller than the strong force, the weak force is responsible for the nuclear reactions in the sun and other stars.

Quarks are subject to the strong nuclear force whereas leptons are not. Both quarks and leptons have the electric and the weak charges, and both are affected by gravity.

Now the Standard Model (first formulated in the 1970's) is certainly not the last word on the fundamental structure of the physical world. As particle theorist Gordon Kane observes, the Model can't explain new phenomena such as the "dark energy" and "dark matter" believed to pervade the universe or the fields that supposedly drove the inflationary expansion of the universe. Neither does it unify gravity with the other three forces. Nevertheless, Kane points out, new discoveries and theories are likely to be extensions of this Model rather than replacements. The two leading extension candidates are the Minimal Supersymmetric Standard Model and M–Theory (described further below).

Will the Real Theory of Everything Please Stand Up?

String theory is the best known of theories that attempt to unite all known particles and forces. String theory tells us that quantum field theory applies at low energies. The fundamental constituents of the material world, according to this theory, are not zero dimension particles but one-dimension string-like entities; particles are excitations of these strings. These strings vibrate in different ways and thereby take on the characteristics of fundamental particles, e.g., electrons and photons, and forces; additionally, they are about the trillionth of a trillionth size of an atom. Further refinements of string theory postulate the existence of strings with different dimensions called branes ranging from two-dimensional membranes to nine-dimensional p-branes. The five competing string theories have now been harmonized under the umbrella of M–Theory (M doesn't stand for anything definite; some proponents take it to refer to Membrane, others to Magic).

M–Theory and its predecessors postulate the existence of several more than the three spatial dimensions of our experience. There are various views on the number of the other spatial dimensions (10, 11, 26) and their size.

As even M–Theory's proponents admit, there is currently no experimental evidence for strings, membranes or extra dimensions. No one has observed or detected a string, a membrane or an additional dimension. Recent experiments indicate that gravity travels at the same speed as light. Some commentators have said that these findings do not bode well for theories of extra dimensions, since some of these theories had claimed that gravity could appear to move faster than light by moving through other branes.

Nevertheless, M–Theory shows promise as a solution for fundamental problems:

♦ the dichotomy between general relativity and quantum physics

♦ why there is more matter than antimatter in the universe

♦ why particle masses and force charges have the values they do

♦ the existence of dark matter.

The theory has also inspired a body of mathematical theory that ties gravity, quantum mechanics and subatomic forces and particles into one formalism. It has been enthusiastically endorsed because, if validated, it could help resolve several major conundrums.

But there is the very real possibility that string and brane theories may turn out to be mistaken. Physicists once accepted the existence of phlogiston and the ether simply on the basis of conjecture and built elaborate edifices of explanation around these concepts. In string/brane theory we see much fruitful conjecture coupled with scant empirical evidence. The mathematical physicist Roger Penrose calls it a long shot for this reason.

Whatever the verdict it receives in the long run, M–Theory epitomizes the mind's conviction that the world is built on underlying symmetry and unification. It's also a testimony to the mystery and intricacy of the world of mass/energy that we experience day to day.

Real-World Applications
Like all the other major milestones in science, the quantum breakthrough has had significant real-world applications in science and technology.

Additionally, quantum physics has shed light on virtually every field of science. As noted above, it is especially relevant in any study of the four fundamental forces of nature, since each force results from the emission of an exchange particle by one subatomic particle and its absorption by another. The structure of each element, and thus of the periodic table, is again understood in terms of quanta and their interactions. Moreover, matter is cohesive at the subatomic level because the *entangled* (see entanglement above) state of the electrons ensures the strength of their bonding.

With regard to technology, arguably no other scientific theory has spawned as many applications as quantum mechanics. In fact, a significant portion of the GNP of most nations is today generated through quantum-related technologies. Modern inventions that owe their birth to quantum physics include the transistor, television, lasers, semiconductors, superconductors, cell phones, laptops and new forms of computing and communication. Nuclear power is one of the more awesome displays of quantum phenomena, since radioactive decay is a quantum mechanical process.

> *Geek*: I think there's little here that's questionable. But it's the interpretation of the theory that's always been the sore point, at least for some. The more ardent Copenhagists even claim that the foundations of quantum theory—mathematics, models, and experiments—are inextricably embedded in their particular interpretation.

Guru: You're right about the claim. But I hope to show that things are a lot more complicated than they make it out to be.

4. What Quantum Physics Tells Us About The World

The puzzling phenomena uncovered by quantum investigators have led to even more puzzling philosophical dilemmas. In fact, the most notorious paradoxes in quantum theory are not scientific but philosophical. But many quantum theorists have expressed their opinions about these meta-scientific issues without seeming to recognize that they are doing philosophy and not science. There is nothing wrong with scientists indulging in philosophy, but when they do so, they must give philosophical arguments and not appeal to the authority of science. As I pointed out earlier, under cover of the quantum, philosophies and fallacies that died a long time ago have reincarnated themselves. These include Idealism, Positivism and even solipsism, the idea that nothing exists other than oneself. Many of these philosophical positions have been held by thinkers of the past, and have died well-deserved deaths.

But the data of quantum physics do not lend support to these positions, just as the principles of classical physics cannot be said to prove the opposing view. In the first place, all scientific theories are in principle revisable, and so they cannot serve as reliable starting-points for a philosophy. Secondly, by their very nature, philosophical claims are dramatically different from scientific hypotheses.

So what part of quantum physics is scientific and what philosophical? When we deal with physical parameters or quantitative data, measurement or observation, we are doing science. When we talk of reality or a law of cause-and-effect or the nature of human knowledge or something coming from nothing, we have wandered into metaphysics. The physicist as physicist can legitimately say that it's not possible to simultaneously measure momentum and position at the quantum level. But a physicist who concludes from this that nothing is real until it is measured is speaking as a philosopher and making a philosophical claim.

In my view, we have to address certain key philosophical issues if we wish to make extrapolations from quantum theory:

♦ The distinction between fact and interpretation, science and philosophy

♦ Solipsism and observer dependency

♦ Fundamental unknowability and fundamental intelligibility

♦ Cause and effect

♦ Something coming to be from nothing

♦ The need for experiments to verify scientific claims and mathematics to describe them

The three most significant interpretations of quantum physics are the Copenhagen Interpretation of Bohr, Heisenberg and Pauli; the Many Worlds hypothesis of Hugh Everett and David Deutsch; and the Hidden Variables view of Einstein, Schrödinger, de Broglie and David Bohm. The Copenhagen Interpretation (CI) is by far the most influential of the three, and for most students of the subject it is, in fact, identified with quantum theory as a whole. But, without question, it is a philosophical thesis and must be assessed on philosophical grounds.

Bohr belonged to the tradition of Immanuel Kant, who claimed that we can only know appearances and not things in themselves. Bohr's observation that physics tells us not *about* nature but what we can *say about* nature has also been compared to the Logical Positivist idea that science uncovers what *can be said* but not what *can be known*. Biographers have noted that Bohr developed his philosophical outlook before he took to physics, and it was this outlook that served as the foundation of the Copenhagen Interpretation. Thus Bohr's views of what is knowable were not inspired by his scientific work; rather, his philosophical views shaped his interpretation of the scientific data. Kant's Idealist philosophy had a tremendous influence on European thought, and in some respects the Copenhagists were his intellectual descendants.

The focus of the Copenhagen school was on the act of observation and measurement. As its adherents saw it, all experimentation at the quantum realm was inextricably linked to the measuring apparatus. Rather than the actual objects of inquiry, the focus was on the interaction between these objects and the instruments used to measure them. Thus a quantum particle has no property distinct from the measuring device. The ideas of electron spin or momentum, like all descriptions, are specifically related to the specific measuring device that is used. Even the question of whether it is seen as a particle or a wave depends on the device deployed, and clearly no one device can show it to be both.

This wave-particle distinction is a perfect illustration of Bohr's idea of Complementarity, which he believed lay at the heart of the quantum world. As he presented it, although wave and particle descriptions are mutually exclusive, they are complementary, not contradictory; they are simply classical descriptions of non-classical phenomena. There is no independent, objective reality above and beyond individual measurements with specific devices. The results we get will be driven by the kind of instrument we use, and all those using the same instruments and procedures will get the same results. We can't know what is the case, only what we ourselves *make* to be the case. It's simply meaningless to think of a subatomic system independent of the instruments with which it's observed. The quantum realm, in principle, can't be known in itself. As Bohr noted, there is no quantum world as such, just an abstract description of it. He admitted the paradox of using macroscopic instruments to generate definite data on subatomic phenomena, but saw this as a starting point that could not itself be explained.

The Many Worlds hypothesis is in some respects an offshoot of the Copenhagen Interpretation. Copenhagen says that prior to measurement a quantum particle is in many states at the same time and is forced into a specific state only when it is measured. The wavefunction that specifies the various possible states of a particle collapses into one when we observe it and measure its energy or position. Hugh Everett, the pioneer of the Many Worlds view, believed that there is no collapse of the wavefunction, and that all quantum states are equally real. Thus the wavefunction is literally true, and the particle is in one place and in every possible place in other universes. A new universe comes into existence for each state and splits off from our current universe. The most prominent proponent of this view today is David Deutsch who believes that various versions of us exist in parallel universes!

The Copenhagen comeback to Many Worlds theorists is a restatement of their original position: quantum physics is not really about the world but about our interactions with the world. The so-called collapse of the wavefunction is not a literal occurrence but simply a change in our state of knowledge. Thus there is no need to postulate multiple universes. According to Bell of Bell's Theorem, the multiplication of universes is an extravagance that serves no real purpose in the theory and can be dropped. Jeremy Butterfield observes that the space of the wavefunction is configuration space, a mathematical representation, and not ordinary three-dimensional space. The multiplication of universes to realize all possible configurations of particles is therefore grotesque. John Archibald Wheeler, who inspired the Many Worlds interpretation, eventually rejected it because it required too much metaphysical baggage.

Of course, the Many Worlds theory is considerably handicapped by the fact that there is no evidence whatsoever to support it! Any available evidence is tucked away in parallel universes that are necessarily outside our reach. At one extreme, Copenhagen eliminates the one world we have. At the other, Many Worlds creates an infinite number of worlds.

The third alternative is the Hidden Variable hypothesis. Theories in this class have tried to eliminate some of the paradoxes of quantum theory by postulating the existence of certain variables, as yet undiscovered, that would yield a complete and accurate description of physical systems even at the quantum level. The two best-known proponents of this view were Albert Einstein and David Bohm. It is now generally acknowledged that most hidden variable theories have been experimentally disconfirmed, with Bell's Theorem for instance. But the Bohmian position, with particles guided by pilot waves, still remains a contender as a plausible interpretation of the quantum world. It is, nevertheless, a minority view.

In considering these interpretations, let us review the facts that should serve as our starting points and that are, in fact, presupposed by quantum theory (although many quantum theorists seem not to be aware of this).

The Existence of the World

♦   There are three basic issues raised by the Copenhagen view: Does the world exist independent of observation and measurement? Can we know anything about the world? Is there a law of cause and effect in nature? The Copenhagen answers to these questions have been ambiguous. The Nobelist Eugene Wigner once remarked that the Copenhagen people are so clever with language that even after they answer your question you don't know if the answer was yes or no. But the fact of the matter is that neither physics nor any of the other sciences is possible if the answer to any one of the three questions is no.

♦ Copenhagenists like to say that in physics we can only talk about nature, not find out how nature is. As Bohr said, there is no quantum world, only a description of it. The great probability theorist E.T. Jaynes had a superb response: the primary experimental fact of all is the fact of a real world that exists independent of our imagination and obeys its own laws independent of our thought. We receive new evidence for this every waking minute and, as he put it, there is no point pursuing physics or science if this is not the case.

♦ In this context, the most devastating criticism of the Copenhagen Interpretation is the charge of solipsism, the discredited belief that nothing in the world exists but oneself. Because of its emphasis on the role of the observer, Copenhagen invited obvious questions: Does something come into being only when we observe it? Is everything dependent on our knowing it to exist? Does the world exist only if you observe it? Einstein once commented that the moon exists even when we don't observe it. Quantum interactions take place all the time even when they're not observed or measured. It's one thing to say that they can't be precisely defined until they're measured, quite another to say that they don't exist when they're not observed. These are questions about being and non-being, potentiality and actuality, which can't be answered with quips or evasions. Either the world exists or it doesn't. We can't prove the world exists from science, since any experiment in science presupposes its existence. Realization of the world's existence, we have seen earlier, is a fundamental pre-philosophical, pre-scientific insight.

What We Can Know of the World

♦   If we grant that the world exists, the next question is what we can know of it. It's a mistake to assume that the only knowledge we have of the world comes from scientific observation and measurement. What's primary is not the knowledge of the world we get from our senses, important though this may be, but what has been called the intellectual apprehension of the world. Although the senses and the imagination certainly play a key role, in the final analysis our grasp of the world comes from the understanding we have of it. *To perceive something is to understand it with the intellect, whether it is a table or a theorem.* By pointing to the enigmatic behavior of subatomic particles, quantum theory increases our understanding and knowledge of the microworld. But it's not the kind of knowledge that can be pictured by the imagination. Quantum theory tells us that there's no direct fit between our sensory powers and the subatomic realm. Nevertheless the universe is intelligible even at that level in the language of mathematics. Although we're still dealing with quantitative data, what is required is an exercise of the intellect, not the imagination.

♦   There are certain things we do know about the quantum world: we *know* that it's built on constants like Planck's constant and the charge of the electron, and that the laws applicable to it are statistical and probabilistic. We know that our way of knowing is different; the ideas and patterns that we intuitively imbibe in the classical world are replaced with elaborate mathematical formalisms, formalisms that have their own sets of rules. Also, to know that observation/measurement causes the collapse of the wavefunction, as Copenhagen affirms, or that a photon pair from the same source exhibits entanglement is to know something.

♦   Although the Copenhagenists have sometimes seemed obsessed with telling us what we do not and cannot know, the focus in recent years has been on information-relevant possibilities in the quantum realm, specifically in quantum information theory and quantum computing. Christopher Fuchs points out that information processing and transfer is different at the quantum level. The new principles being discovered here enable new kinds of communication and higher computing power. E.T. Jaynes notes that all science, classical and quantum, involves the acquisition and processing of information.

Cause and Effect

♦ The popular idea that quantum theory has eliminated cause and effect from the microworld is just wrong. It can, of course, be said that there are no precisely measurable classical deterministic causes at the quantum level. But this is not the same thing as saying there are no causes. As Peter Hodgson notes, we may not be able to predict which atom is going to undergo radioactive decay, but the decay in the case of each atom is driven by internal parameters, e.g., the state of motion of the nucleons at a particular instant, and quantum mechanical laws. That quantum phenomena follow laws is shown from the fact that atoms of uranium and radium undergoing alpha decay consistently turn into intermediate isotopes and then elements like lead, not into puddles of water or plumes of smoke. An early commentator on quantum theory pointed out the root of the error in *Nature*: the inability to ascertain the cause of a change does not justify the conclusion that the change has no cause.

♦ Quantum physics has shown that the question of what can be known about nature in terms of predicting events and the question of cause-and-effect are different issues. "The limits of predictive calculability of future events have indeed turned out to be in principle incalculable," writes Grete Hermann. "Yet there is no course of events for which no causes could be found in the framework of the quantum mechanical formalism." Because "quantum mechanism presupposes and calls upon an explanation based on natural law also for events which are not predictively calculable," says Hermann, "Gapless causality is not only consistent with quantum mechanism, but is demonstrably presupposed by it."

♦ The Nobel Prize winner Julius Schwinger observed that both classical and quantum physics assume the law of cause and effect, since they both hold that knowledge of the state of a system at one time gives knowledge of its state at a later time.

♦ In any case, the principle that every phenomenon and event has an explanation is as fundamental in quantum physics as it is in the rest of physics. It is not simply a presupposition of science or a thesis to be proven, it is rather a condition that has to be accepted if we are to do science. Like it or not, the quantum physicist cannot say anything about a quantum state that is not implicitly an attempt to explain it. Even to say that there are no causes at the quantum level is to give an explanation, albeit a mistaken one, for quantum phenomena. When Bohr offered his Principle of Complementarity, he was trying to explain quantum phenomena. It could be said that this was a description not an explanation, but then it was a description that sought to explain. All quantum experiments and formalisms are attempts to explain quantum phenomena. Quantum theorists are often, in fact, driven by the search for symmetry. Bizarre quantum phenomena like non-locality may be bizarre from the viewpoint of conventional science. But the non-locality experiments actually re-introduce causality; we do not understand how one photon affects the other at such distances but we do know that one has an effect on the other.

♦   Now cause-and-effect is actually a meta-scientific principle. Science cannot prove it; science simply operates on the assumption that it is valid. In the Einstein-Copenhagen debate on measurement and causality, Copenhagen (Bohr) had the right premise but the wrong conclusion. Einstein adopted Copenhagen's conclusion as his premise and reached an equally flawed conclusion. From the premise that exact measurement is not possible at the quantum level, Copenhagen concluded that the law of cause and effect does not apply there. This process of reasoning is obviously flawed because the existence of cause and effect between two phenomena does not require us to believe that this relationship can be demonstrated by the rules of classical physics. Einstein, on the other hand, started off with the premise that there can be no cause and effect if there is no exact measurement and concluded that acceptance of indeterminacy at the quantum level eliminates causality and objectivity. Here he fell into the Copenhagen trap because he took exact observation, measurement and prediction as the criteria for cause-and-effect and reality.

Something from Nothing?

♦   Some popularizers have said that quantum theory calls for particles to come into existence from nothing. The basis of such claims is simply ignorance of what the theory itself says. The philosopher

Adolf Grunbaum, who is an atheist, observes that the so-called pair creation of a particle and antiparticle occurs through the conversion of other forms of energy, and he emphatically notes that this is not a creation out of nothing.

Particle and Wave?

♦ Finally, we wonder how a reality can be both a particle and a wave. Robert Spitzer suggests that there is a simpler quantum state that does not have the boundaries of either a wave or a particle. In certain settings, this simpler state takes on particle-like boundaries and behaves like a particle; in other settings it has wave-like boundaries and acts like a wave.

As you can see, I'm not in any sense negatively inclined towards quantum physics. If anything, I am far more positive that some of its prominent practitioners. Let me end by listing the three features of the quantum world that I find most fascinating in the context of our discussion:

♦ The incredible complexity at the most fundamental level of the world is startling. We've barely scratched the surface of this unthinkably rich universe, and have already reaped bountiful rewards in our understanding of the rest of science and the development of revolutionary technologies.

♦ Despite the intrinsically unpredictable nature of the basic building blocks of the world, everything surprisingly settles down into orderly, precise structures at the level of the macroverse. How does this transformation take place? The difference between classical and quantum physics simply mirrors the radical difference between the micro and the macro worlds.

♦ How is it that certain precise constants reign in the quantum realm? Life became possible only because these constants had the exact values they did.

We've considered the structures and laws of the macroverse and the microverse. Inevitably these led to questions about the origin of the universe. A systematic treatment of this issue seems therefore the logical next step.

Geek: For a non-scientist, I think you're doing quite well so far.

You've hit the hot buttons with panache. I'm really curious about your take on modern cosmology.

Guru: I can't wait to see what you think of my approach in this area. That's next.

Chapter

The Big Bang and Before

*H*ow did the universe originate? Most people are familiar with the theory that all of space and time, matter and energy, came into being approximately 13.7 billion years ago in the event known as the Big Bang. Current histories of the universe and of life on earth take this event as their starting-point. When people ask what happened before the Big Bang the standard answer used to be that time itself began with this event, and so the question of "before" made no sense—it's a question like "what is north of the north pole?"

In the last two decades, cosmologists and theoretical physicists have proposed revisions to the notion that the universe and time actually began with the Big Bang. All kinds of adaptations, extensions and outright transformations of the original theory have spawned exotic breeds of cosmologies. These theories are built on such brand new concepts as:

♦ Fluctuations of the vacuum and inflation of one or more universes

♦ Quantum Gravity with such ideas as imaginary time (already reviewed)

♦ A multiverse made up of universes generated from quantum energy fields, black holes or the multiple parallel histories of every particle

♦ Cycles of expansion and contraction that have no beginning or end in which our universe springs into being from the collision of parallel membranes in multi- dimensional space

To reiterate the point that needs to be repeatedly made, almost all the what-happened-before-the-Big-Bang models of cosmology are built on speculation not fact, theory not experiment, ideas not data. This point needs to be underscored because proponents often fail to admit that their theories have a speculative rather than factual basis. Some progress was made when the eminent cosmologist P. J. E. Peebles graded popular theories in cosmology in *Scientific American*:

♦ The evolution of the universe from a hotter denser state got an A+,

♦ The expansion of the universe as predicted by General Relativity received an A-,

♦ The theory that dark matter dominates galaxies got a B+

♦ The acceleration of the universe got a B-.

The theory that the universe grew out of inflation got an "Incomplete" because, in Peebles' view, it lacked direct evidence and required a huge extrapolation from the laws of physics, although it's certainly compatible with the data from recent space probes.

This evaluation ties in with the distinctions I drew among the four levels of evidence: hard facts established beyond reasonable doubt; reasonable hypotheses rooted in circumstantial evidence; plausible speculation without any physical evidence; and implausible speculation that lacks evidence and flies in the face of what we know. The more fundamental distinction between science and meta-scientific ontology is especially relevant in cosmology. The primary meta-scientific insight that will guide my inquiry here is the absolute truth that something cannot come into being from nothingness. Ignorance of this principle is the primary source of confusion in many cosmological discussions. Many cosmologists use the term "nothing" when they mean a kind of something.

To make any progress we've got to separate the wheat of evidence from the chaff of speculation.

Right at the start, let me tell you that I do not see the Big Bang as proof that God created the universe, let alone think of it as the Creation Event. The reason for all the mayhem in modern cosmology is this: no one seems to realize the rules of the game when it comes to ultimate origins.

There is only one important question when it comes to the fundamental origin of all things: why is it the case that something exists instead of nothingness? For centuries, skeptics were able to shrug off this question by declaring their faith in an infinite static universe that always existed. The Big Bang woke such smug skeptics from their slumber. One point became clear: things weren't as static as they thought. When the initial shock of this revelation wore off, they regrouped their forces to reclaim the high ground of a dynamic universe, with the added proviso that it was endlessly so. Thus many modern theories go behind the Big Bang to posit eternal inflation or invoke the mystically multiplying multiverse or recycle the idea of a cyclic universe.

But none of these theories come to grips with the fundamental issue of how it turns out to be the case that something exists. If at any point absolute nothingness was all there was, then there's no conceivable way in which any universe or anything at all would ever exist. But if there existed some set of laws that determined the evolution of one or more universes, then we have to ask what is the source of these laws. But this is a question I would like to address after reviewing the raw data of modern cosmology.

1. The Basis of Big Bang Theory

Our starting point is the Big Bang theory in its elemental form. Although the theory, viewed as a whole, is not free of anomalies, certain of its key elements have strong experimental support.

As I see it, the Big Bang theory paints a picture of seven different eras in cosmic history:

+ Era of primordial unity and perfect symmetry
+ Era of force differentiation, when the forces of nature separate from each other
+ Era of particle formation
+ Era of nuclear formation

- ◆ Era of atom formation
- ◆ Era of light/matter differentiation
- ◆ Era of galaxy formation

There are four things I find remarkable about this picture. Most people think the novelty of the Big Bang theory lies in the way it explains the origin of the Universe. Just as important, however, are these other implications of the theory:

- ◆ A universe made up of hundreds of billions of stars and galaxies existed in its earliest stage as a subatomic-sized particle.

- ◆ For the first 300,000+ years of its life, the universe was nothing but GAS and radiation.

- ◆ All of the elements and chemicals and materials with which we are familiar were born in a cosmic game of "Survivor."

- ◆ Right from the start the universe has been impelled by an unbelievable burst of energy, the origin of which is just as puzzling as the origin of the universe as a whole.

Just as amazing as this picture of the universe is the human ingenuity that made it possible for us to discover it. We have created tools that take us back in time and across space:

- ◆ Telescopes on earth and in space that show us the young universe by detecting light from objects that are millions of light-years away. NASA's Wilkinson Microwave Anisotropy Probe (WMAP), for instance, is in orbit a million miles from earth.

- ◆ Particle accelerators that recreate the formation of the elements and the high-energy state of the early universe.

- ◆ Satellites and balloons that detect the cosmic microwave background radiation generated by the initial expansion.

We are indebted also to the intrepid astronauts who have probed the world beyond our planet at great personal risk, sometimes sacrificing their own lives, as happened most recently with the Columbia disaster.

Modern scientists reached the conclusion that the Universe began its history in a hot dense state on the basis of observational and experimental data accumulated from the 1920s through the 2000s. At a given instant, some 13.7 billion years ago, give or take 200 million years, an infinitesimally small quantum state of being, billions of times smaller than a hydrogen atom, underwent a phenomenally rapid expansion that produced the colossal cosmos we inhabit today. This expansion event is the phenomenon popularly termed the Big Bang.

Three hard facts have emerged in cosmic origin studies along with four more speculative ideas that may or may not be verified definitively. The hard facts are:

- the expansion of the universe
- its cooling
- the nucleosynthesis of the elements that constitute it.

The speculative ideas are:

- Quantum Gravity from the era when the universe was a quantum particle
- the symmetry-breaking of the fundamental forces of nature
- the inflationary universe
- the apparent acceleration of the universe, supported by some experimental evidence.

There are also the more esoteric domains of colliding branes and bubble universes that again in principle cannot be verified.

The phrase "Big Bang," coined by an antagonist of the theory, is misleading. There was no explosion involved because an explosion implies bits and pieces flying out into space, whereas the Big Bang was the event that created space. As Einstein showed, it is the distribution of mass/energy that creates space, and the Big Bang is the event that distributed all existing mass/energy, starting from the instant at which all of it was concentrated in a primordial point of the highest possible density. As this all-encompassing point expanded and bodies of matter moved away from each other, space was formed. Space, in the specific sense of conduits where motion is possible, came to be at the same time as the galaxies. From a point of the highest density, the average density of the universe continues to decrease as the expansion proceeds and drives clumps of matter further apart.

The expansion of the universe is sometimes compared to the baking of a loaf of bread or the inflation of a balloon. If you think of space as the dough, galaxies as the raisins on top and baking as the expansion, then the rising of the dough automatically moves the raisins further apart. Alternatively, consider a balloon with dots on it. As it is inflated the dots are driven further apart, not because the dots themselves are expanding but because the rubber surface is being stretched apart. In the balloon model, it is evident that there is neither a center nor an edge to the universe.

The events that proved decisive for the cosmos as a whole took place in the first minute after the Big Bang. The particle interactions that took place then determined the structure of the larger-scale universe. Although the two earliest periods in the history of the universe, the hypothetical Quantum Gravity and inflationary eras, remain in the realm of speculation, scientists have been able to plausibly reconstruct the subsequent chain of events—an astonishing achievement. They have done this by studying the rate of expansion of the universe, the amount of matter in it, the observable radiation, and the make-up of the elements.

Although there are a few variations in the details of these estimates, the general outlines of the timeline detailed below are fairly well accepted (see, for instance, Hawking's *The Universe in a Nutshell*). The time units are plotted relative to the time of the Big Bang. If the Big Bang took place at t=0, all other events took place after this initial point.

☙

Time	Era	Evidence	Events
t=0 to 10⁻⁴³ secs after the Big Bang	*Quantum Gravity Era*	Sheer Speculation	According to prevailing cosmological models, an ineffable explosion, trillions of degrees in temperature, created not only fundamental subatomic particles and thus matter and energy, but space and time itself. The universe in its entirety was initially compressed into a single infinitely dense point the size of an atomic nucleus. The first 10^{-43} seconds after the Big Bang event, a period referred to as Planck time because it is the shortest possible interval of time, has been called the Quantum Gravity era. The universe was ostensibly so small and so dense that none of the conventional laws of physics, either quantum or classical, applied. Various theories of the laws governing the universe in this state and of what happened before have been floated but there is no realistic way to determine the truth of any of these views. No evidence is available and there is no experimental procedure that could verify the validity of something that took place only once. Although Planck time itself lies beyond what can be known, physicists have used known physical laws to describe the subsequent evolution of the universe.

| 10^{-43} to 10^{-35} secs after t=0 | Grand Unification Era | Speculation based on behavior of the evolving universe | During this era of enormous temperatures and massive densities, fundamental forces and particles take center stage. *Gravity breaks away from the three other fundamental forces. Matter triumphs over antimatter.* Energy in the form of photons (packets of light) is dominant. These photons collide with each other to produce matter and antimatter, e.g., quarks and anti-quarks. Particle pairs are formed from heat energy (thermal photons). Quarks and leptons are the fundamental building blocks of matter. Particles and antiparticles annihilate each other releasing energy. Although initially there are an equal amount of quarks and anti-quarks, by the end of this period there is a tiny surplus of quarks. For every thirty million anti-quarks, there are thirty million and one quarks. Thus one quark survives every time thirty million quarks and anti-quarks destroy each other, and this incredibly small proportion of surviving quarks constitutes all the matter in the universe. Quarks today are always housed within larger particles like protons and neutrons but because of the high temperatures of this period, they are free, i.e. not confined. The excess of matter over antimatter is a puzzle because it seems inconsistent with the conservation laws of nature, but if not for this excess, we would not exist. The background radiation of the seething cauldron of photons constitutes most of the energy density of the early universe. The proportion of matter made up of the surviving particles, is comparatively very small. |

10⁻³⁵ to 10⁻¹⁰ sec	*Electroweak Era*	Circumstantial evidence. Inflationary theory is speculation at best.	*The strong nuclear force splits from the electroweak force.* According to inflationary theory, in this period, a few brief instants after the Big Bang, there is an especially "fast" expansion of the universe. Some theorists believe that the symmetry breaking of the nuclear and electroweak forces and the accompanying phase transition triggered this inflation. The expansion of space has a cooling effect. Whereas heat energy created particles, cooling enables these particles to stick together. The quarks will soon condense into amalgams called hadrons, namely, protons, neutrons and the like. Thus photons and quarks and leptons are to be joined by protons and neutrons.
10⁻¹⁰ to 1 sec	*Hadron Era*		*The end of symmetry: the electroweak force divides into the electromagnetic and weak forces.* The temperature at this time is 100 billion degrees. With the drop in temperature, in a process termed baryogenesis, quarks are confined in composite particles: baryons (three quarks) and mesons (two quarks). The most stable baryons, also called hadrons, are protons and neutrons. Neither mesons nor unstable baryons exist today. Most of the universe is made up of lighter particles, photons, electrons and neutrinos, with a smaller number of protons and neutrons. Protons and neutrons turn into one another, but the weak interaction between them keeps them in balance.

1 sec to 3 min		Data from high-energy and nuclear physics lab and from observation of the universe	*Protons and neutrons stick together to form the nuclei of the light elements, hydrogen, helium, lithium and deuterium.* The temperature is now 10 billion degrees. As the universe expands and thereby cools and decreases in density, there is a corresponding decline in particle interaction (since temperature is a measure of the energy per particle). Neutrinos decouple from matter. The formation of electron-positron pairs comes to a halt because there is less energy for photons to form them. The initial parity in the transformation of protons and neutrons into each other ceases because it takes less energy to create a proton than the heavier neutron—and the weak force can no longer maintain balance in view of the rapidity of expansion. Soon there are seven protons for every neutron.
			An element is described in terms of the number of protons in its nucleus. Hydrogen is the simplest element with only one proton. After the formation of protons and neutrons, the only element in the universe was hydrogen. As the universe cools in the first three minutes to 1 billion degrees, protons and neutrons are able to bind together to form nuclei in a process known as nucleosynthesis.
			The first compound nuclei to be produced from this nuclear fusion are the light elements: deuterium with 1 proton and 1 neutron, helium-3 with 2 protons and 1 neutron, helium-4 with 2 protons and 2 neutrons, and lithium with 3 protons and 4 neutrons. Approximately 23% of matter ends up as helium with tiny traces of helium-3, deuterium and lithium and the rest remaining as hydrogen. There is a clear correlation between the predictions of the proportions of the light elements in the early universe and the actual relative abundance of these elements.

300,000-500,000 years	Recombination Era		*Electrons combine with nuclei to form atoms. Matter decouples from radiation and the universe becomes transparent.*
			Approximately 300-500,000 years after the Big Bang, the temperature of the universe has dropped down to about 3000 degrees. Until then the vigorous collisions driven by high temperatures prevented electrons from binding with nuclei to form neutral atoms. Protons and nuclei zigged and zagged through an opaque soup of photons, electrons and neutrons. It was opaque because photons could only travel short distances before they ran into other particles. Matter was electrically charged and interacted with light. Likewise, electrons trying to bind with nuclei were torn asunder by photons.
			But once temperatures had dropped to 3000 degrees, photons no longer had the energy to separate electrons from nuclei and protons. The energetic electrons were captured by nuclei to form atoms, and matter itself was now neutral. Unhindered by the electrons, and thus freed from matter, the photons expanded with the universe and filled it uniformly with radiation. At the end of this period, the energy density of matter surpassed that of radiation.
			The cosmic microwave background radiation first detected in the 20th century comes to us from this era 300-500,000 years after the Big Bang. Some 13.7 billion years after the Big Bang, the temperature of the radiation is a mere 2.73 degrees Kelvin. The variations in intensity of this radiation are indications of the seeds of future galaxies.

1 billion years			In the next billion years, atoms are pulled together by gravity to form gas galaxies that gradually condense into stars and proto-galaxies. WMAP indicates that the first stars ignited 200 million years after the Big Bang. Over billions of years, helium and hydrogen are cooked at the center of stars to produce heavier elements like carbon and oxygen. As the stars explode, these elements are flung through space and then condense to form stars and planets. Life appears some 9-10 billion years later.

The Big Bang theory has been proposed to explain certain hard facts. Although the theory has been disputed, these facts are generally accepted. The three hard facts, as mentioned before, are the cooling and expanding of the universe and the nucleosynthesis of its elements.

Expansion

One of the major breakthroughs of the twentieth century was Edwin Hubble's deduction that the universe is expanding. This deduction was based on a study of the light reaching us from distant galaxies. Waves of every kind have a specific effect depending on whether or not we're approaching or moving away from them. In the case of sound waves, for instance, the pitch of a train whistle changes as it approaches us or moves away. The same applies to light waves. A source of light moving toward us has shorter wavelengths and settles on the blue end of the spectrum, while the light moving away has longer wavelengths and comes to rest on the red end.

Galaxies transmit or absorb different wavelengths of light in different degrees depending on position and velocity. A galaxy expanding away from earth will have light of longer wavelengths and therefore a red shift. Observations of starlight from galaxies far away show it to be systematically shifted to the red end of the spectrum. This red shift implies that these galaxies are moving away from earth and that space is expanding. Hubble showed that the red shift is proportional to distance and thus the red shift of a galaxy closer to earth is lesser than that of one further away. According to Hubble's law, the velocity of a galaxy moving away is equal to its distance from earth multiplied by a number called Hubble's constant. The farther away a galaxy is from the earth the faster it is moving away. If we extrapolate backwards in time, assuming that velocity remained constant, we conclude from the observed expansion that the galaxies in each region were closer together and the universe itself was denser.

Remarkably, the observation that the universe is expanding came a few years after it received a theoretical foundation in Einstein's General Theory of Relativity, which implied that the universe would have to either expand or contract.

There are three questions about the expansion of the universe for which scientists are trying to find answers: What triggered it? Why does it appear to have the same rate in all directions, and this over a period of billions of years? And how did it come to have a rate that is just right? If it were just slightly faster, stars could not have formed and if it was just a little slower, it would have contracted before there was time for the stars to form.

Cooling

The discovery that the universe was expanding was followed by the equally significant observation that it was cooling down. The Big Bang story begins with all the galaxies we observe today being packed into an infinitesimally tiny point. Two things follow if this really was the starting-point: the universe then would have been very hot because compression produces heat and even today there should be some remnant of that radiation. It should get hotter and denser the further you go back in time.

As far back as 1948 scientists predicted that the remnant radiation would now have a temperature of 5 degrees Kelvin, which is 5 degrees over absolute zero, -273 Celsius. In 1965, two scientists at Bell Labs, Arno Penzias and Robert Wilson, accidentally discovered what is called the Cosmic Microwave Background Radiation, the radiation that comes to us from the earliest eras of the universe. It was found that the radiation had cooled to 2.73 K after billions of years of expansion. Satellites and balloons today monitor the CMBR. Its temperature has been observed to be the same in all directions, indicating the uniformity of the early universe. The famous COBE (Cosmic Background Explorer) study of 1992 found that there are certain variations in the temperature, 1 part in 100,000 and these tiny inhomogeneities may indicate the fluctuations in energy and density of the early universe that were amplified by gravity to form galaxies. WMAP provided higher resolution readings of temperature fluctuations to within a millionth of a degree of the CMBR. Analyses of these new readings have helped provide answers to at least some key questions.

Nucleosynthesis

The Big Bang is known from its fruits that exist to this day. All the particles and elements that comprise the physical world were "cooked" in three stages:

1. First, in the gigantic particle accelerator that was the Big Bang;
2. Next, through the subsequent cooling of the universe;
3. Finally, the nuclear burning of stars and the subsequent dispersal of the elements like carbon and oxygen formed therein; supernova nucleosynthesis, as distinct from normal nuclear burning, gives heavier elements.

Of the three stages, we have a tentative idea of the recipe in the first stage, a somewhat fuzzy idea of the second and a clearer picture of the last.

Like the primordial radiation, the products of the nuclear reactions that took place at the time of the Big Bang should, if the theory is true, still be around today. The post-Bang universe was initially almost entirely made up of hydrogen (75%) and helium with traces of lithium and deuterium. Studies of the chemical makeup of the universe thus far have confirmed that the lighter elements are present in proportions consistent with the theory. It is hydrogen that fuels all the stars in the universe powering up sunlight and starlight.

There are other key questions in cosmology for which there are still no clear answers. These can be answered, if they can be answered at all, only through continued investigation: What laws governed the universe when it was point-sized? How are the four forces of nature related to each other? How much matter and energy is there in the universe, and what is its density? What proportion of the universe is made up of matter and energy that cannot be directly observed? Will the universe keep expanding or will it contract at some point? Promising discoveries in these and other areas, such as the data generated by WMAP, are frequently announced. But a brief overview of the issues will show that some answers are necessarily tentative in nature while others lie outside the pale of observational data.

Quantum Gravity

If the universe started off as a speck, the question is what laws applied to it at its earliest stage. This period is called the Planck Era because all quantities then were at the Planck scale. The size of the universe, 10^{-33} cm, and the interval of time in which it existed in this state, 10^{-43} seconds, were the smallest possible multiples, beyond which division is not viable. So which laws apply at this level? The laws of the universe as a whole or the laws of its tiniest components? Is it the Theory of General Relativity that applies to the large-scale universe or quantum laws that apply at the subatomic level? The problem is that neither theory has been reconciled with the other. Relativity deals with gravity and space-time, quantum physics with subatomic forces and particles. Since the universe at this stage is measured in quantum sizes and distances while also subsisting with the greatest possible density and energy, it seems that only a quantum mechanical theory of gravitation can describe its properties. In other words, relativity and quantum theory have to be unified under a new framework that has been called Quantum Gravity.

There is presently no established theory of Quantum Gravity, although a few hypotheses have been offered. The two most prominent attempts to find a solution are M–Theory or string theory and Loop Quantum Gravity. The first starts off from quantum theory while the second takes relativity as its starting-point. Currently neither theory has been experimentally tested or verified. Since empirical confirmation is the ultimate determinant of fact in science, we cannot treat either theory as established. Nevertheless the only hope of a solution to the problem of Quantum Gravity lies in ceaseless probing and creative theorizing.

Symmetry-Breaking

Three classes of theories are associated with the four forces of nature:

1. Theories about the unification of all four forces are called Theories of Everything.

2. Grand Unified Theories (GUTs) are concerned with the unification of the strong, weak and electromagnetic forces.

3. The Standard Electroweak Model proposes the unification of the electromagnetic and the weak forces.

As yet there is experimental evidence only for the Electroweak Model and some circumstantial evidence for a GUT.

It is hypothesized that all the forces of nature were one at the start of the Big Bang. This symmetry was broken only when the universe started cooling because the so-called superforce dissociates into various forces at lower energies. Gravity first diverged from the three other forces, then the strong nuclear force split from the remaining two, and finally the weak separated from the electromagnetic force.

These symmetry-breaking events are called phase transitions because matter changes its phase, e.g., its form and properties. We witness these transitions in daily experience as water goes through various phases at different temperatures: it turns into ice when frozen and becomes steam when boiled. The way in which particles interact is affected by the energy available to them. The particles are most symmetrical at higher temperatures; steam is more symmetrical than water and water more than ice. It was no different in the early universe.

Changes in temperature affected the interaction of energy. What was originally one integrated force at very high temperatures, i.e., higher energies, disintegrated into four as the temperature came down. If this truly happened, in theory these forces should merge again and particles would return to their previous states if the higher temperatures are replicated. Although it is impractical to recreate Big Bang-type temperatures, scientists have succeeded in producing the three vector bosons that carry the weak force by recreating conditions of the universe when it was about 10^{-12} seconds old and the temperature was 10^{15} Kelvin. This experimental evidence helped confirm the electroweak theory about the merger of the electromagnetic and the weak nuclear forces. Production of the exchange particles mediating the merger of the electroweak force with the strong nuclear force would take a temperature of about 10^{29} Kelvin. This would require a particle accelerator the size of the solar system. The reunion of all four forces would require exponentially higher temperatures.

☙

Dark Matter, Dark Energy, the Density and Destiny of the Universe

It used to be said that there were only two numbers that were important in cosmology: Hubble's Constant, which is the velocity of expansion of the universe, and Omega, the ratio of the amount of matter in the universe to the amount required to halt its expansion(its critical density). Three scenarios are possible depending on the value of Omega. If Omega is more than 1, then the expansion will halt at some point and the universe falls in on itself in a Big Crunch (the closed universe). If it's less than 1, the universe continues to expand forever unhindered by matter or gravitation (the open universe). If it's 1, the universe is exactly balanced between collapse and expansion (the flat universe).

So how much matter is there in the universe? Studies of the motion of galaxies indicate that most of the matter in the universe is hidden. This hidden mass is called dark matter because it doesn't emit or absorb light. It is known through its gravitational effects. We don't know what makes up this dark matter. Various candidates, from black holes, brown dwarfs and neutron stars to massive neutrinos, axions and particles from the Big Bang era, have been proposed. But they remain hypotheses. Many cosmologists believe that Omega = 1, with most of the mass being made up of dark matter and dark energy.

Dark energy is one of the newer additions to modern cosmology. Astronomers studying the distant dazzling explosions that signal the death of a star discovered that these supernovae looked dimmer than they would be if standard models were right. This has led several theorists to the conclusion that a repulsive force was speeding up the expansion of the universe. The force may well be a kind of dark energy that makes up most of the universe. The acceleration hypothesis has given rise to a number of follow-on ideas:

♦ Various observations, e.g. of the microwave background radiation, and calculations have led cosmologists to assume that space is flat not curved, so that light travels in straight lines. But this view runs into the parallel finding that there is not enough matter, both visible and dark, in the universe to make space flat. The existence of a dark energy, however, would provide the missing ingredient required to account for the shape of space.

♦ Its apparent acceleration has given rise to predictions that the universe will eventually end up as a sea of black holes that disintegrate into particles and then finally decay into a void. According to Fred Adams and Greg Laughlin, this will only happen in one trillion, trillion, trillion, trillion, trillion, trillion years time. Instead of collapsing in on itself, the universe will keep expanding into oblivion.

Cosmologists don't know what constitutes the dark energy. Nor do they know its properties or the value of lambda, the energy of the quantum vacuum. Is dark energy like the repulsive force counteracting gravity—called the cosmological constant—that Einstein arbitrarily introduced into his equations and then retracted? Is it a force that changes with time, so that at some point it could become attractive, like gravity, instead of repulsive?

Some theorists have used this apparent phenomenon of acceleration to develop new cosmologies. This is a risky strategy because the evidence is tentative and could be overturned. For instance, it has been said that gravitational lensing, a distortion of light that produces multiple images of the same galaxy, would be more frequently observed than it actually is if we live in an accelerating universe. Counter-explanations have been offered, e.g., the repulsive force varies in different times and places, but it's clear that the case for acceleration is not airtight.

Answers to some but not all of the questions above were offered in February 2003 by NASA's latest Wilkinson's Microwave Anisotropic Probe (WMAP), which has been studying the so-called afterglow of the Big Bang, the light present 380,000 years after the Big Bang, since June 2001. A study of temperature variations in this radiation enabled scientists to get a glimpse of the universe in its earliest stages, before the formation of galaxies and other structures. Below are some of WMAP's findings:

- The Universe is approximately 13.7 billion years old, with a 1% margin of error.

- The first stars ignited 200 million years after the Big Bang.

- The Universe is made of 4% matter (atoms), 23% cold dark matter and 73% dark energy.

- The dark energy is characterized as a kind of cosmological constant.

- The results of the probe are consistent with certain versions of inflationary theory, i.e. "the predictions of the most basic inflationary models are in good agreement with the data."

♦ The universe will continue expanding instead of collapsing in a big crunch.

♦ The universe is flat, i.e., space is neither positively nor negatively curved.

"The emerging standard model of cosmology," note the scientists who collaborated on the probe, "fits the WMAP data." So what is this model? "Over the past century, a standard cosmological model has emerged: With relatively few parameters, the model describes the evolution of the universe and astronomical observations on scales ranging from a few to thousands of megaparsecs. In this model the universe is spatially flat, homogeneous and isotropic on large scales, composed of radiation, ordinary matter (electrons, protons, neutrons and neutrinos), non-baryonic cold dark matter, and dark energy. Galaxies and large-scale structure grew gravitationally from tiny, nearly scale-invariant adiabatic Gaussian fluctuations. The Wilkinson Microwave Anisotropy Probe (WMAP) data offer a demanding quantitative test of this model."

The authors end with an overview of what is known and what still needs to be found:

♦ "Cosmology now has a standard model: a flat universe composed of matter, baryons and vacuum energy with a nearly scale-invariant spectrum of primordial fluctuations."

♦ "WMAP's detection of TE [temperature polarization] fluctuations has confirmed the basic model."

♦ "Cosmology is now in a similar stage in its intellectual development to particle physics three decades ago when particle physicists converged on the current standard model. The standard model of particle physics fits a wide range of data, but does not answer many fundamental questions: 'what is the origin of mass? Why is there more than one family?', etc.' Similarly, the standard cosmological model has many deep, open questions: 'what is the dark energy? What is the dark matter? What is the physical model behind inflation (or something like inflation)?' Over the past three decades, precision tests have confirmed the standard model of particle physics and searched for distinctive signatures of the natural extension of the standard model: supersymmetry. Over the coming years, improving CMB, large scale structure, lensing and supernova data will provide ever more rigorous tests of the cosmological standard model and search for new physics beyond the standard model."

2. From Whimper to Bang and Back

The experimentalists in cosmology continue to generate new and exciting data on a daily basis. Meanwhile, the theorists have been hard at work generating new and exciting cosmologies that present, so to speak, the big picture. If there was a Big Bang, what caused it? And what happened before?

Whereas the experimentalists rely on telescopes and satellites, lasers and computers, laboratories and accelerators, the theorists work with mathematical models and innovative concepts. Sometimes the theorists are able to use raw data but by and large they have to go far beyond the available data. Of necessity their visions tend to be shaped by their own metaphysical and aesthetic preferences. I will give a brief overview of some currently popular cosmologies and then drill down to the fundamental issues that require an answer. The key cosmological views summarized are inflation and in particular chaotic inflation and the cyclic theory of brane collision.

Inflation

Inflation is an idea that was introduced to resolve several glitches in the classical Big Bang model. Given that this model calls for a super-hot, immeasurably dense beginning and a rapid expansion, how is it that the universe today is so smooth? The expansion made it improbable that matter and radiation could be distributed uniformly since this meant that information and energy would have to be transmitted a hundred times faster than light. The paradox is partially resolved if you posit an exponential expansion a trillionth of a second after the Big Bang when the universe grew in size by a factor of 10^{30}, i.e., one thousand billion billion billion times.

In this scenario a small region suddenly turns into the entire universe, and the uniformity of temperature in that region automatically carries over into the universe it becomes. Inflation also explains the existence of galaxies because minor quantum fluctuations in the small region allowed galaxies to form at larger scales. The apparent flatness of space is also a consequence of inflation; as the universe keeps doubling in size its curvature is straightened out.

So what caused the inflation? In most popular theories, an energy field causes the expansion of space and then disappears leaving the standard hot soup of particles in its wake. The immediate trigger in the initial version is what is called a false vacuum. According to Alan Guth, the best-known proponent of inflationary theory, a true vacuum is a state of the lowest possible energy density while a false vacuum has a very high energy density. It is false because it is temporary and it's like a vacuum because its energy density cannot be rapidly lowered. False vacuums are unstable and inflation is caused by this instability. Very high temperatures and extraordinary energies led to a false vacuum with gravitational repulsion, instead of attraction. The false vacuum was a billion times tinier than a proton; in some versions, a hundred billion times smaller. It was all that existed at the time of the Big Bang. The gravitational repulsive force of this false vacuum, says Guth, caused a constant doubling of the universe every 10^{-37} seconds. This doubling continued until the false vacuum decayed into a soup of particles; its energy is converted into particles. The appearance of mass/energy did not violate the law of conservation of mass/energy because the energy of gravitation is negative and balanced the positive energy of the new mass. Guth even hypothesizes that portions of the false vacuum may decay at random to produce new bubble universes.

Andrei Linde and Alexander Vilenkin, two other major proponents of inflationary theory, have made this extended inflation the centerpiece of their theory of "the eternally existing, self-reproducing inflationary universe." Some regions of the false vacuum decay into universes, while others continue to expand. This is a phenomenon Linde calls "chaotic inflation" and scalar fields, energy fields like electrical or magnetic fields but with no direction, drive it. In Linde's version, inflationary universes produce new inflationary domains that then produce new domains and on and on indefinitely. Linde says the universe is immortal inasmuch as new universes keep appearing, but he is not sure about the beginning. Although every part of the universe began with a singularity, he says there is no evidence as to how the process started. The obvious question is whether all these parts were created from a simultaneous general singularity. While holding that the universe eternally re-creates itself, he admits that this does not solve the problem of the issue of the initial creation, except that it pushes it back to the indefinite past. Linde says that the Big Bang is now a subset of inflationary theory rather than the other way round.

What evidence is there for inflationary theory? Guth says that the theory accounts for the size of the universe, its uniformity, its apparent flatness, the fine-tuning of its mass density, the formation of the galaxies and the way in which galaxies move away from each other, the Hubble constant. Inflationary theory, with its notion of repulsive gravity, is also compatible with the apparent evidence that the expansion of the universe may be ac-celerating. The acceleration is said to show gravity acting repulsively as required by the theory. Critics, as we shall see, have a different view. Thus the theory seems to resolve a number of anomalies in the classical Big Bang model. This is by no means definitive proof. The fact of the matter is that the false vacuum has never been observed. Moreover, there are now about fifty different inflationary theories, old, new, chaotic, extended, hybrid inflation, to name a few, with different accounts of how the four forces are unified and different rates of inflation.

Inflationary theorists have said that the discovery of gravitational waves would help establish their view. Heisenberg's Uncertainty Principle says that energy fields constantly create virtual particles that then annihilate each other. During inflation, it is theorized that virtual gravitons, the hy-pothetical carriers of the gravitational force, were wrenched away before they could be annihilated and their wavelengths were stretched out. In principle, it should be possible to detect these gravitational waves from the Big Bang era since they affect the regions through which they travel, if they exist that is. But scientists like Martin Rees do not believe that either NASA's WMAP or the Planck satellite to be launched by the European Space Agency in 2007 will be able to detect the polarization associated with gravitational waves. Nevertheless, the WMAP researchers have noted, "WMAP both confirms the basic tenets of the inflationary paradigm and begins to quantitatively test inflationary models." They add that "the detection and measurement of the gravity-wave power spectrum would provide the next important key test of inflation."

❧

Cyclic Universe

But inflation is not the only current view of origins. Two pioneers of inflationary theory, Paul Steinhardt and Neil Turok, now reject the theory and propose a radically different alternative. Their endless cycle theory does away with both the Big Bang and inflationary theory and has its roots in M–Theory, previously string theory, with its idea of multiple dimensions called branes.

The entire theory revolves around one piece of very tentative data: the indication that the expansion of the universe may be accelerating, an acceleration apparently driven by dark energy. Just as the observation of the expansion of the universe was explained by reference to General Relativity, the theorists in this instance have tried to explain the acceleration data by appealing to string theory. It hardly needs to be said that Relativity is a very well established theory while string theory is still a creative mathematical exercise.

The theorists frankly admit that they are reinvigorating ancient cosmic mythologies with a cyclic view of the universe in which there's an endless sequence of contraction and expansion. Space and time, in their view, have always existed. The history of the universe is an endless series of Big Bangs and Big Crunches in which each Bang is followed by a Crunch. There are trillions of years of evolution between each Bang and Crunch, in which the universe accelerates slowly and empties its entropy and black holes, triggering off a contraction. The universe keeps bouncing back from negative to positive energy without settling in at the ground state; in this model the vacuum energy of the ground state is not zero but negative. There's no inflation, no infinite density or temperature. A major flaw in previous cyclic theories was the problem of explaining how a Crunch could lead to a Bang since the laws of physics would not allow such a transition. The new theory invokes M–Theory to get around this problem.

Cyclic theorists say there is only one universe that is made up of a hidden five-dimensional space-time with multiple branes (short for membranes) embedded in it, like sheets of paper in our regular three dimensional space. Our universe is an infinitely large brane at one end of the five-dimensional space. The cosmic expansion was triggered off when we collided with a brane at the other end that has different laws of physics from us. The kinetic energy of the collision was transformed into the particles and forces of our world. After trillions of years of expansion and cooling, the two branes will collide again, a process that goes on without end. The dark energy that drives cosmic acceleration is a leak of energy in the form of gravity that passes between the two branes across the fifth dimension.

The cyclic theorists highlight what they claim are major flaws in inflationary theory. It is alleged that dark energy and cosmic acceleration cannot be explained by inflationary theory, whereas these are crucial ingredients in the cyclic theory. And while inflationary theory predicts two periods of accelerated expansion, in the early universe and then again in current acceleration, the cyclic view requires only one such period. Secondly, inflation assumes that different regions of the universe are in different stages of inflation, whereas in the cyclic model each region of the universe is typical of the universe as a whole. Thirdly, the cyclic view, it is said, offers a more complete picture of cosmic history than inflation, which simply talks of events after the Big Bang.

Inflationary theorists have responded with scathing critiques. Linde, for instance, says the cyclic model is a complicated theory that doesn't work. For good measure, he adds that it's a bad idea popular only among journalists. Cosmologists in general are skeptical. Critics have noted that the fatal flaw in theories of endless cycles is the physics required to bounce a contracting universe back into an expansion mode. Whether M–Theory can provide the bridge is yet to be proven since M–Theory is far from established.

One is at a loss in knowing how to respond to theories of this nature that can neither be verified nor refuted by any known experimental data. Although cyclic theorists admit that the discovery of gravitational waves would be fatal, they offer no positive evidence in favor of brane-collision, an idea that is fundamental to their theory. The cyclic proposal is at best a metaphysical theory with some mathematical models that describe it; these are descriptions and not proofs. Although it purports to explain dark energy, the explanation itself is dependent on another highly speculative theory, M–Theory. In any case, as a metaphysical view, the theory does not even begin to address the question of why there is something and not sheer nothingness. It simply teaches that there is an endless cycle of expansion and contraction without offering any proof for the teaching or offering an explanation of how there came to be an ensemble with these properties. Turok, one of the proponents of the cyclic theory, says the universe is the way it is because it has to be this way in order for it to repeat the next time around!

Once and Future Theories

Indubitably, newer even more outlandish theories of how the universe(s) came to be will be proposed. In the past, several such theories appeared with a bang and vanished without a whimper.

The most famous such casualty was the Steady State theory. The theory, advanced by Fred Hoyle, Hermann Bondi and Thomas Gold in 1946, maintained that the universe was infinite in space and time. While accepting the expansion and homogeneity of the cosmos, the theory made the notorious claim that new matter is created out of nothing as the universe expands. This preserved a constant density of matter (steady state). The theory died because it could not cope with the scientific data. The discovery of the cosmic microwave background radiation, in particular, dealt it a deathblow because the theory did not allow for a high-density hot state and the resultant radiation. Philosophically, it was dead on arrival because it made the unbelievable assumption that matter pops out of nothingness on cue.

The oscillating universe hypothesis held that the universe goes through endless cycles of expansion and collapse. This hypothesis is a cyclic view of the universe in which oscillation is the mechanism for recurrence. The problem with this view, highlighted earlier, was the impossibility of showing through physical law that a universe in collapse could bounce back. We cannot even show how matter sucked into a black hole could bounce back, let alone a universe that contracts into a singularity. And a major proponent of the theory, Richard Tolman, admitted that the theory had no answer to the question of how the process got started, in other words, where did the initial energy come from. The larger question is how did a universe with oscillating properties come to be.

Another model of an eternal universe was the plasma cosmology of Hannes Alfven. In this model, the universe, which is made up of electrically conducting gases, eternally recycles energy. Alfven rejects the Big Bang partially for philosophical reasons. In his view it implies divine creation, an option he finds unattractive. Plasma cosmology has had a difficult time finding observational data to support its predictions, and much of the available evidence is in conflict with its claims. Hence it has not had much of a following. But even if it were true, it still leaves open the question of how things got started. The theory holds that the universe started with uniform hydrogen plasma. At the same time it admits it has no knowledge of what processes brought the plasma into being.

The recurrence of the cyclic model in cosmology is matched by the appearance under different guises of the multiverse theory, the notion that there are an infinite number of universes. One variant of it was the Many Worlds interpretation of quantum physics. Every quantum event in the cosmos, like the collapse of the wavefunction of a particle, creates multiple new universes. A new universe is created for every possible outcome of the quantum event. These new universes themselves split into other universes and all the universes evolve in parallel. There's no way to prove the theory since no physical evidence is possible in principle; no contact is possible between any of the universes. Another evidence-less multiverse idea is the theory that black holes create baby universes because of their enormous energies, and these also evolve in parallel. Again, we cannot reach out and touch these universes since it's impossible to make it in and out of a black hole. The latest multiverse idea, we have seen, is the bubble universe model of the inflationary theorists. In Linde's model, universes spring out of quantum energy-fields at random everywhere and all the time. Astronomers like Martin Rees have speculated that the apparent fine-tuning in the universe exists because at least one of the infinitely many universes will have the particular constants and conditions that made life possible. He admits, however, that the scientific case for a multitude of universes lies on the speculative fringe of cosmology; the idea is built on guesses not laws or evidence.

Certainly there's no prohibition on theorists concocting theories that can't be scientifically verified immediately. At that point the theories become metaphysical and should be judged as philosophies. But if a theory is to qualify as scientific it can't wander too far from the experimental data. The great astrophysicist Robert Jastrow, an experimentalist of the highest order, once said that theoretical physicists sometimes let their mathematics run wild while anchoring it to very little in the way of observation. Another experimentalist, Saul Perlmutter, notes that theorists tend to go overboard for whatever is the latest new idea, whereas experimenters like him tend to assume the universe is very complicated and each new item of information shows how much we don't know.

There is one thing all multiverse theories share in common: there's no physical/empirical evidence available to prove them. Neither are there any established laws of physics that indicate their existence. It's a purely speculative play. It's less science than it's science fiction. The scientist James

Trefil notes, "The reason that the notion of parallel universes seems to go in and out of fashion is that cosmologists never construct theories to deal solely with this phenomenon. They are always trying to adjust a theory to produce the right mass density for the universe, for example, or the right proportion of dark matter. In each case, the question of whether or not the theory also predicts parallel universes is something of an afterthought." In a sense, he concludes, "whether or not we are thinking about parallel universes at a given time is an accidental consequence of the prevailing cosmological theories."

Even if we grant the existence of parallel universes, we're still no closer to answering the question we started with: why is there something and not sheer nothingness? Where did this ensemble of universes come from? If it is from fundamental laws and processes that incarnate themselves in a primordial field, how did those laws or the field with its potentialities originate? The Princeton cosmologist J. Richard Gott, who talks of a mother universe from which other universes emerge, says the mother universe, which is sustained by energy from the quantum world, creates itself and makes the first matter *in some way we will never be able to know*. We're back to square one. Alan Guth is well known for his comment that the universe is the ultimate free lunch. Inflation shows that everything can be created from nothing, he says, and then adds "or at least from very little;" on other occasions he has said, "well, almost nothing."

The common thread in all the theories we have surveyed is a tendency to substitute physical facts with metaphysical conjectures. There is nothing inherently wrong with metaphysics. But metaphysics that contradicts self-evident truths is as wrong-headed as a scientific theory that contradicts hard facts. Clearly, what's self-evident to one person may not seem so to another. But what we call self-evident truths are the principles that are presupposed by science. These can't be verified by science but we can't do science without assuming their truth. For instance, we assume that there's an explanation for every phenomenon without claiming that we can discover the explanation. If a metaphysical theory rejects this assumption, it must first justify its rejection before it can be taken seriously.

In the final analysis, any hypothesis about the *ultimate* origin and/or nature of the universe, or multiverse if you prefer, that goes beyond experimental evidence, has to address the issue of whether or not it has an ultimate explanation, an explanation that tells us why there is something instead of nothing. If you say the universe always existed, then you still need to explain the phenomenon of an eternally existing universe. How did it come to be that there is a universe with this property of endless existence? The only viable explanation, in my view, is a source of existence that has no limitation.

> *Geek*: I must say I'm somewhat surprised you're not resting your case for a Creator on the Big Bang. But is this wise from your viewpoint? Since there are several plausible explanations for the existence of the universe on a purely natural level, you seem to be left with Hawking's conundrum: there's nothing for a Creator to do. This is of a piece with Laplace's observation that he has no need of the hypothesis of a God to make his science work.

Guru: Actually, the central thrust of my exposition goes in the opposite direction.

First, science can't tell us anything of a Creator or supernatural interventions. Science can deal with data that are observable, measurable and quantifiable. It has tools and methods to generate and structure these data. It furnishes models and paradigms with which we interpret the data, all on a natural level.

Second, science makes certain assumptions that can't be verified by scientific methods. Some of these principles are fundamental to all scientific work, for instance the law of cause and effect or the idea that our minds can understand the world and how it works. We are free to reject these assumptions. But, if we do so, we can't consistently work as scientists or accept the data of science. If we say there's no relation between cause and effect, then we can't accept any law of science. If we say that the world is unintelligible then we can't propose any scientific theory, let alone accept one.

Third, and most important, science can only deal with something and not with nothingness. On the purely scientific level, we have a lot of fascinating data that tell us about the current state of the universe along with fairly plausible hypotheses about its previous states. But in considering the previous-most state of the universe, science bumps up against a barrier that's insuperable. It can't observe and measure either nothingness or the transition from nothing to something.

Paradoxically enough, the strongest arguments in favor of my affirmation of infinite Intelligence, God, are the very arguments used by those who deny a meta-scientific explanation for the origin of the universe. Let me explain:

- Hawking seeks to show that a combination of imaginary time and the wavefunction of the universe will show that there's no singularity, i.e. a breakdown of natural law;

- The inflationary theorists claim that scalar fields can explain everything that's happened in the history of the universe;

- The latest proponents of the cyclic theory of the universe invoke branes and dark energy to avoid singularities.

They're all driven by the conviction that *there is an explanation* for everything. Not only do they believe there's an explanation, but they also seek to show that the explanation conforms to the laws of nature. None of them are saying that the universe or anything in it happens causelessly. The quantum theorists among them might hold that events at the earliest stages were not predictable or determined, but they don't deny that there were certain conditions that had to exist, e.g. a vacuum with a precise structure, for certain effects to be produced. *It is this fundamental conviction that there is an explanation for everything that drives all of science and simultaneously brings us to the Mind of God.*

So here's my basic point. If the existence of everything in the world can be explained by the existence of certain fundamental laws of nature, we are inevitably led to the questions of how there came to be such laws and how these laws produced a universe that follows their dictates. Great scientific minds (Einstein, Hawking, et al.) have cut through all the mere clutter to go right to this fundamental issue. It is interesting that Einstein, Hawking and other scientists thought of these laws in terms of the Mind of God. And this is my central point: both the intelligence manifested in the universe through its laws and the very existence of this intelligent universe point naturally and inescapably to an Intelligence that has no limitation, an Intelligence that shows its face in the splendor and genius of the world.

ɕɞ

3. The Anthropic Principle

Most accounts of the Big Bang Theory today are accompanied by an addendum called the Anthropic Principle. The upshot of the Principle is that the universe is hospitable to life because its basic parameters were fine-tuned from the beginning to permit the appearance of life. For instance, if the expansion rate of the universe was any higher, there would be no time for galaxies to form, and if the force of gravity was any higher, expansion would stop within 30,000 years of the birth of the universe and all matter would fall back on itself. Several inferences are possible from the fact of factors of this kind:

♦ The precise parameters required for life emerged purely by chance in a universe that self-organized into its current state;

♦ There are a large, perhaps infinite, number of universes and every possible variation of mass/energy, of fundamental constants and natural laws would eventually emerge (this is called the multiverse idea);

♦ The intelligence in the universe is itself a product of intelligence, of an infinite Mind that fine-tuned it for life.

I personally believe that the first two options strain credulity. But let's start with the facts on which we all agree before discussing interpretations.

The more that modern cosmology learns about the origin of the universe, the more it becomes clear that its initial conditions and fundamental laws had to have certain specific values and ratios for life to exist. The slightest change in these quantities would have made the appearance of life impossible. In *Just Six Numbers,* Martin Rees identifies six numbers associated with the underlying properties of the universe that had to be exactly what they are for life to exist. If there was the slightest change in any one of them, neither stars nor complex elements would be possible, and life would not be.

Rees' catalog reads thus:

♦ The strength of the force binding the constituents of the nucleus had to be precisely what it is in the helium atom for fusion to take place in the sun in the way it has. We know that hydrogen is the most prevalent element in the universe. We know also that the conversion of hydrogen into helium generates the energy fueling the sun. It is precisely 0.7% of a hydrogen atom's mass that turns into energy in the formation of a helium nucleus. If the percentage had been 0.6% then there would only be hydrogen atoms in the universe. No stars or elements would form, since the bonding of its nuclei would have been too weak. If it had been 0.8% the bonding would have been too strong and there would be no hydrogen available for energy since it would all be used up to configure heavier nuclei. In sum, if the number were .006 instead of .007, there would only be hydrogen in the universe and if it were .008, there would be no hydrogen after the Big Bang.

♦   The strength of the electrical forces holding atoms together is hugely greater than the gravitational force between them; any decrease in the ratio between them would have decreased the life span of the universe and the inhabitants of the world would not have gotten bigger than insects.

♦   If gravity had been any stronger than it is, then the universe would have collapsed before life arose; if it had been any weaker, then galaxies and stars could not form.

♦   If the strength of the force that governs the expansion of the universe were any stronger than it is, then again stars and galaxies could not have formed.

♦   The seeds of the future structure of the universe were sown in the Big Bang. There is a ratio of 1 : 100,000 between two fundamental energies that enable the formation of galaxies and planets. A smaller ratio would have produced an inert universe of gas and a larger one would condense matter into black holes so that stars could not be born.

♦   The three observable spatial dimensions of the universe are again the right number because neither two nor four would permit life.

Rees notes the extraordinary fact that there is no dependence of any one of the numbers on the others, so that you couldn't deduce any single value from any one of the others.

Other authors have noted dozens of other key factors that facilitated life and a few of these are cited here:

- A slight excess of matter over antimatter at the earliest stages of the universe allowed the material substratum required for life.

- Stephen Hawking observes that stars could not have burnt hydrogen and helium if the electron had a different electric charge than it does. Either that or the stars could not have exploded. Both scenarios would have meant no life.

- The proton is exactly 1,836 times heavier than the electron, a proportion required for molecules to form.

- There is a delicate balance between a star's gravitational and electromagnetic forces; if they didn't balance, the stars would be red giants or blue dwarfs.

- There is an optimal right combination of stars that are hot energy sources and cold planets, optimal because the chemical reactions required for life could only take place with this particular blend.

- The creation of carbon from the collision of helium and beryllium was indispensable because life as we know it is carbon-based. Moreover, the chemical constitution of carbon is very important; the energy of the carbon nucleus needs to be exactly what it is for other elements to be formed.

- Any change in the numerical value of fundamental constants, Planck's constant, for instance, or alpha, the strength of the electromagnetic force, would have jeopardized the formation of life. In addition to the well-known constants, there are, says John Barrow, 25 basic constants that govern the masses of elementary particles and their interactions.

So what conclusions are we to draw from these and numerous other fortunate ratios and parameters? It seems to me obvious that the only inference that makes sense of the data is the existence and activity of an infinite Intelligence that is the ground and driver of the finely-tuned fecundity of our universe. At one time it was biology that housed these kinds of design arguments. Today it is physics and chemistry that illuminate at a fundamental level the precariousness and improbability of human life.

There's an important difference between the new and the older approaches. William Paley had claimed that the design in nature calls for a designer much as a watch on a moor suggests a watchmaker. Charles Darwin sought to discredit the appearance of purpose and design at a natural level by supplying the adaptive mechanism of random variation and natural selection. Paley had provided an analogy and Darwin, biologists believe, exposed it as inapplicable.

But the Anthropic argument is not an analogy. Rather it argues from the hard fact of fine-tuning to a Fine Tuner. This is not an argument from analogy, since it doesn't compare the universe to some obvious example of design; it simply points to the obvious intelligence manifesting itself in the universe. Neither is it an argument from probability as often thought. It is essentially one more step in our continuing discovery of the hierarchy of intelligence in existent being. Every major breakthrough in science has expanded our vision of this hierarchy: motion, mass/energy, fields, molecules, DNA, and the brain. The Anthropic Principle tells us that the appearance of intelligence itself followed an intelligent path. Certainly the skeptic can claim that what appears to be an intelligent path is simply an accidental sequence, or the inevitable product of a multiverse. But there is no denying the appearance of intelligence. And it seems simply inane to suggest that the genesis of intelligence could itself be unintelligent.

> *Geek*: But does the Anthropic Principle work as an argument? In one sense it's like Paley, an extrapolation from current experience. It's a teleological argument and is subject to Joseph Silk's critique of teleology as tautology: we are here and therefore our appearance must be part of a plan. If the universe did not have its present properties, we wouldn't be here to talk about it. Obviously no one disagrees that the universe would need the parameters required for life; this is obvious because we do have life. This proves nothing one way or another. It's like saying, "Life appeared because the conditions required for life were present." Well, yes. After something has taken place we can deduce how it took place. But this is not a prediction or a revelation. It's just a glimpse of the obvious. The multiverse explanation is pretty convincing. Inflationary cosmology suggests that we're likely to have multiple universes bubbling out of the vacuum. Each has laws and so-called fundamental constants shaped by initial conditions. In this context it was quite possibly inevitable that a universe such as ours with life-generating conditions would exist.

Guru: As I've already said, the multiverse idea as such doesn't solve the basic problem; it simply moves it back another level. And it's a prime example of extrapolation! Besides, the idea has no evidence to support it and no possibility of every being demonstrated, since we can't ever generate observational evidence in its favor. It's about as far as it's possible to go from a probability-type explanation. It's not for nothing that Richard Swinburne remarked that the postulation of a trillion trillion universes to explain the orderliness of our world is the height of irrationality. While expressing his distaste for the extravagant multiverse hypothesis, Paul Davies adds that an ensemble of universes still requires an ultimate explanation for its existence and its laws: "It is perfectly consistent to believe in both an ensemble of universes and a designer God. Indeed, as I have discussed, plausible world-ensemble theories still require a measure of explanation, such as the law-like character of the universes and why there exists a world-ensemble in the first place." Thus any multiverse idea still leaves open the question of God.

The idea of a random self-organizing universe is similarly not an ultimate explanation since we still want to know why it has its self-organizing properties. The skeptic might say this is an ultimate brute fact, but that's an emotional and not a rational response.

Certain formulations of the Anthropic Principle are obviously wrong-headed. The so-called Strong Anthropic Principle with its claim that intelligent observers had to exist for there to be a universe seems simply muddled. Likewise, I cannot make any sense of the Weak Anthropic Principle that says that the observed universe allows life because it couldn't be observed if there were no observers.

My focus here is on the hard fact that, thanks to a whole array of improbable factors, the universe is hospitable to a hierarchy of intelligence with life at the top. How do you explain this fact? Well, neither multiverse nor randomness are explanations. The fitting response is a comprehensive explanation of the intelligent networks of laws that not only provide the conditions required for life but also appear actively engaged in bringing us into being. As I said earlier, not only is the world rationally ordered but it's also ordered towards the production of rational life. The laws themselves are intelligent and purposive.

To move closer to a comprehensive explanation I will review the story of life and then consider the origin of the laws of nature.

Chapter

DNA and the "Facts" of Life

1. Family Tree

*I*DON'T BELIEVE that we will disagree on the raw data of the history of life in the world: the fossil record and the themes of a struggle for survival, adaptation to environmental conditions and mutations in genetic material. We start with the same body of fact and circumstantial evidence built up by thousands of scientists in the course of the last hundred plus years. For the most part, we should have no differences on the chronology of events: the initial appearance of life 1.9 to 3.85 billion years ago, the pre-Cambrian Era made up of the Archeozic and Proterozic periods of unicellular and some primitive multicellular life (bacteria, protozoa, Ediacarans); the Paleozoic Era starting with the Cambrian period 535 million years ago when every animal group, all thirty of today's complex animal phyla, appeared in a mere 10 million years, followed by the Orodovican, Silurian, Devonian, Carboniferous and Permian periods (plants, fishlike vertebrates, first land animals, amphibians, insects, reptiles); the Mezoic Era with its Triassic, Jurassic and Cretaceous periods (dinosaurs come and go, first mammals, flowering plants, snakes and lizards); the Cenozoic Era with its Paleocene/Eocene, Oligocene and Miocene epochs (modern plants, modern birds, elephants, cats and dogs, primates); and finally the Pliocene and Pleistocene eras with modern humans emerging 150,000 years ago. Over 1.5 million species of animals and plants have been discovered, and it's estimated that another 7-14 million species are yet to be identified.

I'm not about to dispute what specialists in geology, paleontology, genetics, biochemistry, zoology and botany tell us about their findings and insights. In fact I marvel at both the spectacle of life manifesting itself across our planet over millions of years and the perspicacity of the scientists who pieced together this mind-boggling jigsaw puzzle.

The starting-point for me is this body of data, with, of course, the understanding that all scientific models and schemata are in principle revisable and that current hypotheses may be overturned by future facts. However, there are certain things that precede the biological data and that make sense of the data, but which are not themselves a part of the data. What is life? What does it mean to act? What is purpose? What is consciousness and mind? What constitutes the identity of a person? On a purely scientific level, we cannot find answers to these questions no matter how much we study the data or how much data we study.

The scientist as scientist can talk strictly of the data and of models and laws that can be used to structure the data. Thus the biochemist in pursuit of the origin of life can tell us that life perhaps originated in thermal vents on the ocean floor or that the chemical starting-point of life was RNA or that all living things carry their genetic information in DNA and that RNA turns this into protein. But this is not the same thing as telling us what life is at a fundamental irreducible level or what it means to be a living being or what ultimate Matrix explains the very idea of life. These questions lie outside the purview of science just as surely as the question of why something exists instead of there being nothing at all.

> *Geek*: You're well aware that many scientists feel strongly that the data, laws and theories stand just fine on their own two feet. There's no need to appeal to a higher intelligence in order to explain anything. As they see it, ever since Darwin we have shown that (a) life is a purely biochemical phenomenon different in degree but not kind from inorganic matter, (b) the constitution of living beings shows us how, where and when the first life-forms were formed from matter, and (c) over millions of years and under favorable conditions, these first simple life-forms became progres-

sively complex, giving us ultimately the fauna, flora and humanity of the world.

The ancestor of all living things is a primitive cell that was the progenitor of the three domains of life. Both animals and plants derive from eukaryotes and evolved separately in the pre-Cambrian era. Animal evolution starts with marine animals. Invertebrates and protozoans came first. Vertebrates appeared in the Cambrian explosion 530 million years ago and evolved into a vast array of fishes. Some 360 million years ago, these creatures left the water and came on land. They became the ancestors of all tetrapods, terrestrial vertebrates. Mammals appeared some 225 million years ago and dominated life on land when an asteroid struck the earth some 65 million years ago wiping out all dinosaurs.

The whole scientific world now accepts as fact the thesis that natural selection and random mutation over millions of years shaped the complex patterns of adaptation we see in the history of life and everywhere in nature. If you truly accept this picture, as you say you do, let's dispense with your "infinite Intelligence" of the gaps. We have a complete, self-contained, consistent picture where there are no blanks left to be filled.

We've already created viruses, the most basic life-forms, in the lab. It's only a matter of time before we replicate other forms of life. In sum, we've located the body, we've found the suspect standing over it with a smoking gun and we've recreated the crime. If you're determined to remain puzzled about phenomena that have been fully explained, you're creating an artificial problem to make room for a pre-ordained solution.

Guru: I like your last analogy! But, of course, you and I know that the precise picture you've drawn is not quite as tidy as all that.

There are riddles and dilemmas at the foundations of this grand vision.

∽

Beyond Evolution

First, even the most reductionist account of evolution still leaves open larger questions about the origin of the laws of nature and the world as a whole. The philosopher of science Richard Swinburne points out that evolution may serve as a correct explanation for the existence of a complex organism, but it cannot be an ultimate explanation. There could be no evolution by natural selection if there were no laws of chemistry, e.g., inorganic molecules combine to form organic ones that then combine to become organisms, and these are themselves dependent on particular laws of physics. It's the existence of these finely tuned laws that needs to be explained. Responding to Richard Dawkins' claim that Darwin has solved the mystery of our existence, Swinburne observes that at best Darwin shows the universe to be a machine for making animals and humans. This still leaves us with the question of an explanation for the existence and operation of the machine. He writes, "The watch may have been made with the aid of some blind screwdrivers (or even a blind watchmaking machine), but they were guided by a watchmaker with some very clear sight."

At a fundamental level, we're still left with the questions, first, of why there is something and not nothingness and, second, why we're left with this particular something with laws that enable the appearance of life. Evolution calls for adaptation to environment, but where did this ability to adapt come from? And what is the source of the raw material that does the adaptation and of the environment in which it adapts?

Beyond the Fossil Record

Secondly, let's look at the foundational questions raised by the commonly accepted history of life. To begin with, both of us accept the sequence inferred from the fossil record. But after considering the explanatory frameworks built around this sequence, we confront a new dimension of questions.

1. No matter how much the randomist wants to shout the slogan that there's no progress or direction, the fossil record is a record of progression from unicellular to multicellular life, from amoeba to *Homo sapiens*. Whether this end-point was reached accidentally and purposelessly is a question that science cannot answer because science as science cannot discern purpose or goals. You can certainly interpret the data as indicating no purpose. But I find it extraordinary that:

- there is a clear progression
- there are exact matches between sudden leaps forward in the tree of life and the existence of environments that allow these upgrades to survive and thrive
- the matches were made in relatively narrow windows of time.

For instance, you couldn't have air-breathing land dwellers if there was no atmosphere with oxygen. How is it that there was a parallel evolution of both the earth's atmosphere and the air-breathers? Of course you could say that the atmosphere is the chicken and the air-breather the egg, so there's nothing to be surprised about. But we know that the atmosphere didn't cause the existence of the air-breather, so there's no connection of this kind. Of course you might say that I'm simply misunderstanding the evolutionary thesis. According to this thesis, organisms adapt to environmental conditions and the aquatic animal was able to adapt its physiology to survive outside water; this is evolution at work. But my point goes beyond this obvious correlation to ask how there came to be a system in which two parallel sets of accidents, the evolution of the atmosphere and the evolution of animal life, could run into each other, and produce us.

An example might help make my point clearer. Let's say there's a one in a million chance that you'll win the lottery and you just happen to buy a ticket which just happens to hit the jackpot. Although an extraordinary event, this is obviously just a matter of good luck. By now the randomists can't wait to tell us that this analogy supports their approach. They would say that the parallel evolutions of the atmosphere and the air-breather can be construed as the luck of the draw. Extraordinary but easily explicable without invoking any external input. But think again. Our concern is not with explaining how someone was lucky enough to hit the jackpot. Rather, we're asking where the lottery with its jackpot came from. Who wrote the rules; who set up the system where it's possible for a ticket to produce a lucky winner? You're lucky at one level because you won the jackpot but more fundamentally you're lucky because a jackpot and a system for winning it exists.

Thus a narration of the correlation between environment and adaptation does not address the issue of how there came to be a system with both an environment and entities that can adapt to it. When we see the system unfolding so as to end up with a hierarchy of beings ranging from the inanimate to conscious, rational persons, the natural reaction is to see progress here. Now you might say this progression is simply the result of a chain of accidents, but you can hardly deny that there is apparent progress. The question then becomes how there came to be a system that would enable this kind of progress.

2. The story of life, as recorded by the fossils, is a tale of transition and transformation. What is unexplained if you look at just the fossil data is the enabling technology that powered these leaps. Three specific domains stand out.

   (a) How can it be the case that the whole universe of life originated in one unicellular organism? This is a claim on a par with the proposal that the cosmos with its hundreds of billions of galaxies bubbled out of a microscopic quantum vacuum. Strange things have happened, but I have seen no plausible account of how life in all its diversity and plenitude—millions of species, elephants and jaguars, roses and oaks—could at one point have been contained in the confines of a microbe. It's quite possible that an infinitely powerful Intelligence could have chosen to start off small, but short of that it seems to me just impossible. Worse yet, in this particular scheme of things, the original source of all life was not the humble microbe but a collection of inorganic chemicals.

   (b) My next question is: what triggered off the Cambrian Explosion? All the body plans of all living animals were formed in a period of 5-10 million years, a relatively short time in evolutionary history. If it was a question of certain genes being switched on, why did it happen at that particular time and why all of a sudden and why never again? Clearly this phenomenon doesn't comport well with the gradualistic incremental approach. But that's not my point at all. I am just drawing attention to the raw hard fact of incredible transitions and transformations. We might say the mechanism behind them is natural selection and random mutation, but the fossil record doesn't show us either this or any other mechanism. Fossils furnish a record of events, not of the agents behind the events.

   And whatever you might think of their theory of punctuated equilibrium, I think even their critics, e.g., Dawkins, agree that Stephen Jay Gould and Niles Eldredge have shown that there were sudden bursts of speciation in the history of life.

(c) Moving now to what lies beyond science, let's consider those radical transitions that can't be captured by measuring instruments or chemical formulae. Life, consciousness and mind, in my view, can't be explained on a physical plane. Still less can their origin be pinned down, dissected and duplicated by material means. I know this is a controversial view but that doesn't make it any less true. I certainly intend to show why I think this is the case. And by saying that they can't be explained entirely in terms of physics, chemistry and biology, I'm not saying that they're not related to the physical. Nor am I saying there's no explanation for their origin. I for one think that there's a very rational ground for their being.

3. Finally, the idea that the fossil record shows a sequence of simple to complex is actually simplistic. Right from the start, the state of affairs is vastly complex because DNA with its irreducible intelligence is present from the very beginning of life. And it's this same DNA that leaves its fingerprints on every manifestation of life in history. To borrow your analogy, the smoking gun is DNA! And if the fossil record is the body, the suspect is the originator of this life-giving weapon and the environment in which it can work its magic.

> *Geek*: Creationists assume that any discrepancy in the evolutionary story automatically establishes Creationism. Your response hasn't offered any viable theory for the origin of life. I think that, more than anything else, the biochemical basis of life eviscerates the Creationist hypothesis. If life is shown to be purely a physical phenomenon, a complex configuration of matter generated by prebiotic evolution, then every alleged gap is plugged. There's nothing but physics and chemistry at every level. We've made a lot of progress in studying the origin of life, and it has reached a climax with the creation of life in the lab.

Guru: As far as I'm concerned, this is not, and never has been, a debate between Creationism and Evolution. I'm not a creation-scientist just as I'm not a materialist-scientist. I'm an ontologist. My only objective is to differentiate self-evident truths and hard facts from scientific theory and mere speculation.

2. The Nature and Origin of Life

You're on target when you say that life is of foundational importance. As goes life, so goes the rest of the evolutionary picture. The forgotten crux of the whole argument over evolution and creation, materialism and theism, is life, *or more precisely, the question of what constitutes life.* Biology is by its very definition the study of life and this includes the history and nature of all living creatures. But it's a study of the physical manifestation of life, not of its ontology, i.e., of the nature or essence of life. I happen to hold that life is irreducibly different from non-life and propose to show this to be true. By the way, creation of a virus in the lab is NOT the creation of life. A virus is no more a living being than is a crystal or a reactive chemical.

Confusion about life is prevalent because neither evolutionists nor creationists have paid attention to its two dimensions, the physical and the ontological. To be sure there are numerous books and articles and debates on the origin of life, various treatises on the probability of life appearing given random conditions and the like. These are valuable in themselves, giving us new perspectives on the physical template. But hardly any of the writings in this huge body of literature touch on the fundamental question of the nature of life. Among the very few thinkers who've engaged life at this level are Gerald Schroeder, Josef Seifert, E.J. Ambrose, David Berlinski and Paul Davies.

What does it mean to say that something is alive? What is life? Let me start by saying that life is one of those basic realities that can't be defined without reference to itself, just like the ideas of energy or good. In *Origins of Life and Evolution of the Biosphere,* Carol Cleland and Chris Chyba argue that a precise definition of life is not possible at the moment. But life can be understood by describing its characteristics, both physical and ontological. Once we've considered life in terms of its two dimensions, then all of the debates about the origin of life are seen in a new light.

We noted earlier that living things pump out energy, respond to stimuli and replicate themselves. Josef Seifert expands this list to include:

◆　Self-motion that comes from within and not from external laws, unlike subatomic particles or pendulums;

◆　Self-generation through nutrition (metabolism, etc.), growth (morphogenesis), regeneration of cells, procreation;

♦ Sensitivity to the environment;

♦ Irritability;

♦ Adaptive capacities;

♦ Self-regulation;

♦ Integrated wholeness.

A 1993 symposium at Brandeis University titled "What is Life?" ended with one useful definition of life:

> "Life is a succession of energy-producing ELECTRO-CHEMICAL PROCESSES [generated] by a naturally occurring, simple or complex organism composed of a combination of molecules, each consisting of systematically arranged carbon, hydrogen and oxygen atoms, and a few other elements, forming cells, which consume 'food' and produce 'waste,' both consisting of solid, aqueous and gaseous matter; the process is called METABOLISM; the organism is capable of living within its niche or habitat with minimal dependency on other organisms; energy use is manifested by growth with size limits for most; self-healing; possibly movement; self-replication with each offspring slightly different; irritability; capable of modifying their living environment, both beneficially and detrimentally; with eventual termination of energy production, or death. Exceptions are egg, sperm, spore, seed, and virus, which do not consume food and produce waste; the first four are replication structures, and the fifth has premature life-terminating capabilities."

During discussion at the conference it was noted, although not unanimously, that:

1. A free living single cell, such as an amoeba, is alive and therefore exhibits life and

2. A virus, which can exist only within another organism, is not alive and so does not exhibit life.

Above and beyond these descriptions, biological life is primarily and intrinsically a manifestation of intelligence written in the language of DNA (deoxyribonucleic acid). Life is not simply animation but intelligent animation. Centuries ago it was possible to mistakenly speak of its spontaneous generation from matter since life was identified solely with animation. But with the discovery of DNA and its essential role in the mechanism of life and its replication, such naïveté is inexcusable.

Two kinds of organic molecules, proteins and the nucleic acids, DNA and RNA, lie at the chemical basis of life.

We have seen that proteins are made up of strings of amino acids and each cell in the body, other than sex and blood cells, makes about two thousand proteins every second from hundreds of amino acids. In the protein folding process, proteins are assembled from amino acids to form a specific three-dimensional structure that executes pre-programmed functions in the cell. It was pointed out that this process would take billions of years for a supercomputer, a staggering 10^{127} years to be precise, and yet only seconds for living beings.

DNA

Proteins get their marching orders from the molecule called DNA. As the name suggests, DNA is called a nucleic acid because it's found in the nucleus of cells. DNA is a living blueprint for the construction and maintenance of cells. It's the program that determines the sequence of amino acids in every protein. It's the library of genetic information that becomes the vehicle of heredity because it dictates the characteristics of future generations.

Although DNA itself had been discovered in 1869, the structure of the atoms in this molecule was determined only in 1953 by two young researchers, Francis Crick and James Watson. DNA is made from units called nucleotides, each of which has a core component called a base. The biochemical bases that constitute DNA are Guanine, Adenine, Cytosine, and Thymine, usually called G, A, C and T. Guanine and Adenine belong to the family of purines, Cytosine and Thymine to the family of pyrimidines. Genes and chromosomes are made up of molecules of DNA. DNA is described as having a double helix structure because two strands of DNA twine around each other like a spiral staircase, held together by biochemical bonds. These strands are mirror images of each other with hydrogen bonds forming between the bases, for instance between A and T or G and C.

As Ambrose describes it, master copies of the messages encoded in DNA are "photocopied" by another nucleic acid called RNA (ribonucleic acid). Messenger RNAs that perform this transcription function take it to the production line where proteins are being formed. The message from DNA must be decoded or translated for the amino acids, and RNA molecules called transfer RNAs carry out this translation function.

The half-century following the discovery of the structure of DNA has been marked by numerous other breakthroughs:

- ◆ the identification of messenger RNA,

- ◆ the determination of the genetic codes for proteins,

- ◆ the isolation of genes,

- ◆ the creation of methods to sequence DNA ,

- ◆ the birth of genetic fingerprinting, DNA as a unique ID for every individual,

- ◆ the inference that all humans have a common ancestor, mitochondrial Eve,

- ◆ the Human Genome Project.

The intelligent nature of the biochemical constituents of life cannot be overemphasized. Between them DNA, RNA and proteins store information, communicate, construct, synthesize, repair, replicate, and all of this through systematic cooperative, collaborative interaction. The DNA present in each cell contains three and a half billion nucleotide bases, is about two meters in length, and has more information than three complete sets of the *Encyclopedia Britannica*. The English language has 26 letters, the language of computers has two letters (1 and 0) and the language of life, DNA, has four letters. With this language, which allows the formation of a virtually infinite number of genetic combinatorial sequences, DNA spells out the genetic code of every species.

Genes are different configurations of the four bases that make up DNA and carry specific hereditary characteristics. All of the genes that make up the genetic blueprint of human beings comprise the genome of *Homo sapiens*. The Human Genome Project is an attempt to sequence and map all of these genes, thereby unveiling the elaborate blueprint for building every cell. The Project has shown that there are between 30,000 and 40,000 genes in human DNA with sequences of 3 billion chemical base pairs. All the DNA, the genes, in a cell are ordered in structures called chromosomes. With the exception of sperm and egg cells, each human cell has 23 pairs of chromosomes; the sex cells each contain only one set of 23 chromosomes.

All life-forms today share the same genetic material and have the same 20 amino acids in their cells. Scientists have managed to enable a certain strain of bacteria to live with a 21st amino acid, but this isn't something found in nature. The basic template of life, with all the enormous complexity of DNA and RNA, came fully formed billions of years ago with the first appearance of life. The origins of DNA and proteins pose a chicken and egg problem. The formation and replication of DNA requires certain proteins and enzymes, but these proteins and enzymes can't be formed without the blueprints provided by DNA. Much has been made of this discrepancy, but my concern is less with these chemical constraints than with the more fundamental issue of coded intelligence. The mathematician Ian Stewart says that mysterious as the relationship might be between DNA and genes, even more mystifying is the structure of mathematical rules that tells DNA what to do.

In *The Advent of the Algorithm,* David Berlinski has shown that each molecule of DNA is embodied intelligence because it contains information in a set of signs that have their own meaning. It's an algorithm in biochemical code that drives all life with its cycles of transcription, translation and replication. The algorithm transfers information between two sets of symbols, proteins and nucleic acids. Like all codes, what's important is not the carrier of the code, the chemicals, but the message that is carried. Messages imply information and intelligence. But, as Schroeder notes, the chemical laws that describe the bases, sugars and phosphates of DNA can't provide an explanation for its information content or the intelligence driving it.

Intelligent Agents

Its inherent intelligence warns us that there is more to life than its physical manifestations. A living thing, be it a simple organism or an animal, is far more than the sum of its physics and chemistry. This something that is over and above the individual physical parts is not quantitative; it can't be measured or mapped. It's the center of all the action, the moving force, the seat of power. Traditionally it has been called the soul or the principle of life, the psyche of a being. I prefer to describe it as the intelligent agent that executes an information system using what computer buffs might call "machine language," namely, DNA, which in turn directs the hardware infrastructure, e.g. the cells comprising flesh and bones, the energy-rich sugars, chlorophyll, et. al.

We have said that all intelligent agents are purpose-driven. All of their activities focus on goals ranging from nourishment to replication. And these activities are self-directed: the agent is the source and cause of its action. There's no abstract thing we can call life; only individual intelligent agents that are alive.

Now, of course, there are many forms of life, a hierarchy you might say of intelligent agents.

At the most basic level you have unicellular organisms, and these include cyanobacteria and other microbes. At the next level you have plants. We call them intelligent agents as well because in themselves they perform all the actions characteristic of living beings. Tony Trewavas of Edinburgh University holds that the computational capacity of plants is as good as that of many animals, since they acquire information, plan outcomes and use "complex molecular signaling pathways."

Beyond plants there are animals and within the animal world there is, of course, a further hierarchy, from the capacity for locomotion and reproduction to the operation of the senses, instinct, memory. At the top of the ladder of life in the world is the human person who is self-conscious, rational, and free. Each new level of this hierarchy incorporates most of the distinctive features of the one below. But from the lowest to the highest, every kind of intelligent agent has the same basic but sophisticated set of machine language instructions—DNA. Although we have the same instruction-set, there is a wide variety of hardware platforms depending on the kinds of agents involved.

The hierarchy of agents, then, isn't simply to be classified in terms of phylum, genus and species. More fundamentally it can be broken down into the categories of personal and sub-personal. *Homo sapiens* stands in a unique class of personal agents because of its capabilities of conceptual thought and communication, freewill and intentional action, the consciousness of the self, and the awareness of being a person. All other intelligent agents are sub-personal. But there is a hierarchy within the sub-personal realm. Bugs and slime are agents that are lower down the personal/sub-personal scale than animals with higher levels of perceptual consciousness and communication, for instance, cats, dogs and chimpanzees.

I'm pleased to report that at least some scientists have begun to recognize that life is fundamentally agent-based. Stuart Kaufmann, a pioneer of complexity theory at the Santa Fe Institute, thinks living beings should be seen as autonomous agents with home bases in both chemistry and information. Autonomous agents, he notes, are something else besides matter, energy and information.

The Origin of Intelligent Agents

This understanding of life as a matrix of intelligent agency offers a fresh perspective on issues like the origin of life. Like cosmologists tracing the saga of cosmic beginnings, biologists recreating the emergence and diversification of life on earth have much to work with and much to offer. But cosmologists and biologists can only deal with what is observable and quantifiable. They cannot address the ontological with their tools. And the transitions from non-life to life, animation to consciousness, consciousness to mentation, i.e. thought and understanding, are as radical as the transition from nothing to something.

Certainly it's worthwhile finding out when life first appeared and the unlikely but fascinating settings in which it flourishes. Such investigations may tell us:

◆   when the raw materials and environmental conditions required for life first became available,

◆   what physico-chemical reactions preceded and accompanied the genesis of the complex molecular structure of life.

Life did have a beginning and its beginning certainly had biochemical consequences. But this beginning isn't simply a biochemical event; rather, it's a biochemical event driven by the sudden presence of intelligent agents. You might argue that the earliest bacteria weren't intelligent agents but animated products of certain chemical reactions. But the oldest fossils we have of life, going back over 3.5 billion years (or 1.9 billion years if the skeptics are right), are of modern photosynthetic cyanobacteria that eat, excrete, metabolize, move and reproduce. These bacteria meet the criteria of intelligent agency: they are the source and cause of their actions with goal-driven, inbuilt dynamisms of self-generation and information-processing. Science is a study of the quantifiable and the sensory, and neither the nature nor the origin of intelligent agency fall under these categories.

I'm not suggesting here that a molecular biologist studying protein assembly or RNA editing or the geochemist scouting the planet for likely sites of the origin of life should be concerned with questions of intelligent agency. Rather, they should be applauded for the insights they shed on the biochemical machinery of life or the conditions on the early earth that made life possible. In addition to its strictly scientific value, the data furnished by biochemists and other origin of life researchers deepens our sense of wonder as we consider the phenomenon of life.

The Physical Origin of Life

The current account of the origin of life starts with a common genetic ancestor or community of ancestor cells for the three main domains of life: bacteria, archaea or archaebacteria, and eukaryotes. Eukaryotes, the group to which plants, animals, fungi and many unicellular organisms belong, have cells with a real nucleus housing its chromosomes, a cytoskeleton, internal membranes, and mitochondria (plants and algae have chloroplasts). Archae and bacteria are cells without a nucleus but usually with a membrane. Archae and bacteria have distinctively different kinds of lipid and protein structure. It's believed that gene transfers have taken place between the three families of life.

It's generally supposed that life appeared after a crust and an ocean formed on the earth. Conditions on the early earth were far from hospitable, what with erupting volcanoes, comet and asteroid strikes and an atmosphere made up mostly of carbon dioxide with hardly any oxygen. Most of the oxygen in our atmosphere came from the cyanobacteria that first started generating it as a byproduct of photosynthesis 3.5+ billion years back (or 2 billion years before if recent studies are right). Assuming life began over 3.5 billion years ago, there was enough oxygen in the atmosphere some 2.1 billion years ago to enable the survival of aerobic organisms. Remarkably, to this day various forms of life flourish in the most unexpected settings. Researchers have found archae that live 1,500 feet under the earth's crust; rock-eating microbes; and microorganisms that live in the frozen waters of Antarctica and volcanic vents in the ocean. Single-celled archae and bacteria appeared 3.5–3.85 billion years ago, followed by eukaryotes some 1.2 billion years ago and multicellular organisms with differentiated functions 700 million years ago (how unicellular organisms became multi-cellular remains a mystery).

Where did life first appear? There are various views on this matter and, of course, newer ones will continue to be proposed. A popular theory is that it emerged first in the oceans in thermal vents or tidal pools or under the frozen surface. Critics have assailed these ideas on different grounds. It is said, for instance, that the oceans were too salty at the time for life to form. Other settings, such as freshwater ponds or clays and other minerals, have also been proposed.

Could the complex polymer molecules sustaining life have formed at random from chemical units? Francis Crick, the co-discoverer of the structure of DNA, confessed that the likelihood was so small as to be negligible. Consequently he, along with others, re-introduced the old proposal that life on earth first arrived from outer space. But this idea is rarely taken seriously and has been criticized on two primary grounds: (1) the evidence is questionable because meteors coming into the earth's atmosphere would be contaminated by our own elements, and (2) the interplanetary organic compounds that reach earth as debris come in too low a concentration to have generated life on earth. Most important, if life came from interstellar space, we're still left with the question of how it originated.

Life in the Lab
The search for origin of life sites has been accompanied by attempts to create life in the lab. In his famous 1953 experiment, Stanley Miller generated amino acids by sending an electrical discharge through a combination of water, hydrogen, methane and ammonia. It's now believed that the atmosphere of the early earth was mostly made up of carbon dioxide and ammonia and thus this experiment was not relevant to origin of life scenarios. More importantly, the generation of amino acids is a long way away from the generation of life. Even Miller acknowledges today that the real problem is making polymers—like proteins—from simple chemicals like amino acids, something that still eludes experimentalists.

The German chemist Gunter Wachstershauser has shown that some of the chemical reactions involved in metabolism can be replicated with chemicals in the lava pouring out of vents at the bottom of the ocean. In his view, these reactions are the starter pathway of life. It all began with metabolism, which then invented genes and the cells enclosing them. Jeffrey Bada of the Scripps Institution of Oceanography is highly critical of these claims. In the first place, the tests in the lab do not reflect the conditions on the ocean floor; minute amounts of the relevant chemicals are found at the vent, but the lab experiments use enormous quantities. But even more important, a model for the early evolution of metabolism is irrelevant because this takes place long after the origin of life. What should come first are RNA, DNA and proteins. Bada holds therefore that Wachstershauser's work has no relevance to the origin of life.

For many scientists RNA is the starting point, with RNA molecules supposedly being produced by chance and performing the functions of proteins. Experiments have shown RNA enzymes called ribozymes replicating other RNA. But proponents of this view admit that this is not RNA self-replication. They also agree that RNA did not arrive at the beginning of life but took over from another molecule. Critics have argued that the genetic evidence does not favor the RNA view. But the key question is actually this: if RNA is invoked for the creation of proteins, how did RNA itself, which is far more complex than the proteins, come to be?

Some scientists have tried to demonstrate the origin of life through computer simulations. The initiator develops a piece of software that can make copies of itself with the possibility of random changes. These software creatures seem to manifest many of the characteristics of evolution: they mutate and beneficial mutations survive, they compete for memory resources, and they replicate. Proponents say that these cyber-creatures are in fact alive and this is actual evolution. Biologists are rightly skeptical. Silicon chips and electrical impulses are not DNA , RNA and proteins. Lyn Margulis, an influential origin of life researcher, says such replications bear as much similarity to life as dolls to babies.

Now let me return to the theme of creating life in the lab. I say that this is impossible in principle for the same reason that neither the instruments used to perform a Beethoven symphony nor the scores used by the musicians can be thought of as the creators of the composition. Certainly the performance would not be possible without the scores and the instruments, but these are at best complex vehicles for incarnating a code that hides a work of genius. Conceivably, at some point scientists may be able to assemble from primitive sources the ingredients that constitute a basic life-form. In fact, many activities of cells, from replication to transcription, can work in a test-tube under favorable conditions. But such re-creations bear testimony to the ingenuity required to furnish the basic ingredients of life's physical machinery. And, despite promising it for several decades, scientists have not as yet managed to produce a living being from purely material components.

In this context, Mark Shea remarks, "It has always struck me as odd to point to the immense concentration of intellect, will, technology and energy it has taken to do relatively small things in the extremely specialized conditions of the lab (which nature allegedly did without benefit of any of this) and argue that this product of white-hot focus of ultra-controlling human intelligence is clear evidence that absolutely no intelligence was involved in the production of all the rest of the vastly more complex life we see around us. It's like taking years to build a tiny house of cards and then using this feat to say, 'There! This accomplishment shows the Capitol Dome was therefore obviously the product of a hurricane in a marble quarry.'"

I certainly don't believe they will ever produce a life-form in the real-world sense of an intelligent agent that effortlessly processes intricate symbols to generate complex interactions vital for metabolism and replication. You may choose to believe that this will eventually happen, but this would be an act of faith and not a rational inference with any evidence to support it. Paul Davies has said that the phenomenon of the genetic code mediating information between the two languages of life, proteins and nucleic acids, presents a mystery: how can mindless processes set up codes and languages? Living organisms, he points out, have tightly specified complexity and so the question is how meaningful information can emerge spontaneously from incoherent junk.

It was noted that the creation of a virus is not the creation of life. Eckard Wimmer, the creator of the first synthetic virus, in fact, denies that he created life. The virus in his view is just a chemical with a life cycle.

Viruses are made up of nucleic acids, either DNA or RNA, which are active only if they're in a host that's alive. In a host environment, the virus replicates via its DNA/RNA which produce replicas of itself. To the extent that the virus has DNA and RNA, it has a measure of inbuilt intelligence, as do energy fields. But it's not a full-blown agent because it has no independent existence or independent means of replication. It's certainly not an autonomous agent. If you view life as a software program, then a virus can quite literally be seen as a software virus that lives on the program. It can't exist independently. The program must exist first for a virus to be active.

Synthetic viruses have been created in the lab just as synthetic self-replicating molecules, which no one claims is life, have been assembled from pre-existing components. The creators of the virus start with a published genetic blueprint of the virus; synthesize DNA to match this genetic sequence; and then use chemicals to convert the DNA into the viral RNA. The RNA then copies itself and makes the proteins and other constituents of the virus. It's important to remember that a virus made in the lab is made from custom-made DNA and a pre-formulated genetic sequence; this is hardly the same as creating it from scratch. It's the intelligent systematic assembly from pre-existing components of an active chemical configuration.

The Nobel Laureate Werner Arber has said that autoreplication of a macromolecule does not yet represent life. Even a viral particle, he added, is not a life organism, since it only can participate in life processes when it succeeds in becoming part of a living host cell.

Geek: You mentioned Stuart Kaufmann earlier as someone with whom you seemed aligned. Although I don't accept his theory of the origin of life, it's still a plausible account that doesn't require any supernatural intervention. His complexity theory is the view that systems that reach a level of complexity generate a kind of complex order at a natural level. Thus a primitive mix of amino acids and nucleotides would inevitably become a self-replicating, integrated system. He has demonstrated this thesis with a computer model that shows primordial molecules transforming themselves into autocatalytic sets after reaching a level of diversity. These sets are life forms.

Again, in his *A New Kind of Science,* Stephen Wolfram has given an entirely different explanation of the phenomena we've been discussing. In his experiments with computers, he has found that simple computer programs analogous to the simple laws of physics generate complexity without any Designer or Intelligence behind the scenes. Life and all the complexity of Nature automatically emerged from the operation of such simple programs. There is no design or purpose.

Guru: Clearly, biologists have been skeptical of both Kaufmann and Wolfram. With respect to Kaufmann, they've asked for real-world evidence of autocatalytic, self-replicating sets emerging from a soup of this kind. No such evidence has been given. None of the complex chemical systems currently available has generated an autocatalytic set. More fundamentally, complexity theory usually addresses the order emerging in large-scale structures involving multitudes of players, billions of molecules for instance. But life would not be possible without very precise order at the sub-cellular level; the tiniest components of the cell have to coordinate operations with each other for the cell to work at all. And as Paul Davies notes in *The Fifth Miracle,* life is not really an example of self-organization; it is, rather, specified, i.e., genetically determined, organization.

Wolfram assumes that life, consciousness and mind are simply physical and chemical processes. When we discuss consciousness and mind, I will show you why I think this materialist approach is dead wrong. Moreover, if we grant his assumption that the source of all novelty is to be found in the simple computer programs underlying the laws of nature, we will still need to know where these programs come from. Can there be programs without a programmer?

Geek: I'm not sure where you're heading. But your approach is reminiscent of the long-dead philosophy of Vitalism and its idea of a life-force above and beyond physical processes. The discovery of DNA was the death knell of that particular view. There was no need to appeal to some ethereal principle to explain character traits, etc., since it's all written in the genes. Could it be that you're wittingly or unwittingly re-packaging vitalism in the clothes of its archenemy, DNA?

Guru: The answer is no, positively and absolutely no. Vitalism is the theory that living systems contain a non-physical force or energy that plays a causal role; this so-called life-force falls outside physical laws and processes. It was also believed that life is known by intuition, not concept, and reality is organic, not mechanical. Vitalism actually belongs to the same family tree as materialism since it derives from spiritualist monism. And you know what I think about both kinds of monism. I believe in persons, not life-forces, agents, not living systems.

For the sake of clarity, let me distinguish my position from Vitalism. The fundamental reality, as I see it, is intelligence. It manifests itself in both the inanimate and the animate, the mechanical and the organic domains, through laws, processes, capabilities and, finally, agents. Animate beings have their own hierarchy of intelligence ranging from microbes to mammals. Life, consciousness and mind are progressively advanced kinds of intelligence that are irreducible to anything else. This irreducible intelligence is both taken as a starting-point by science and revealed by it.

The origin of life cannot be solved by either the mechanistic or the vitalistic views. I've noted John Maddox's observation in *What Remains to Be Discovered* that life is one of only two outstanding problems in biology to be virtually untouched. The mechanism for the origin of life on earth, he says, remains a mystery despite the evidence that life appeared some 4 billion years ago.

Let's revisit the train of thought introduced by Nobel Laureate Werner Arber:

1. Life begins at the level of a functioning cell;

2. The most primitive cells require hundreds of specific biological macro-molecules;

3. It is a mystery how such already complex structures came together.

We can see why he concludes that a Creator, God, seems to be an adequate solution to this problem. As to how life is introduced in the world, let me get to this after my survey of evolutionary theory.

Chapter

The Biological Blueprint

1. Evolutionary Mechanisms and Purposive Structures

*L*ET ME DO A RECAP at this point. As I see it, the nature of life is the fundamental issue in any discussion of the biological world. You know my position in this domain. There are four other areas germane to our discussion.

1. The existence of purposive structures in the natural world

2. The supposedly exclusive role of natural selection and random mutation

3. The place of Homo sapiens in the hierarchy of life

4. My proposed mechanism of radical novelty in the drama of life

There are three categories of facts at stake in our discussion:

1. *Insights of modern evolutionary theory that are plainly true.* For instance, as I said, although it's by no means fixed and final, I see no reason to deny the chronology of life and the transition sequences of the fossil record that are accepted as standard by the scientific community. The Tree of Life project of the Institute of

Genomic Research will be a giant step forward in interpreting morphological data with molecular genetics. And I certainly don't deny that natural selection and genetic mutation can play a role in the origin of species.

2. *Hypotheses that should be judged on the strength of the empirical evidence in their favor and concerning which different people can reach different conclusions.* Two examples: (a) When and where did life first appear and what were its chemical antecedents on the primitive earth (notice here that this doesn't settle the issue of the nature or ultimate origin of life)? (b) What gave rise to the Cambrian Explosion?

In a paper on the meanings of evolution in *American Scientist,* Keith Stewart Thomson lists the three senses in which the term evolution is understood: change over time, relationship of organisms by descent through common ancestry (he points out that this is a hypothesis not a known fact) and an explanatory mechanism for the first two senses (for instance, the Darwinist model postulates random variation and differential survival at the levels of gene, individual, population and species).

3. *Other claims of certain evolutionists that are plainly false.* These claims are either non-scientific, in the sense that they have no empirical basis, or wildly speculative. Examples of such issues that I call meta-scientific are the nature of life and answers to the four themes outlined above (purposive structures, etc.). I hope to show that my answers to these questions are both self-evident and presupposed by science itself.

Purposive Structures

I want to draw your attention to phenomena that are intrinsically and undeniably purposive and that could never arise from matter per se. In his debate with J.J.C. Smart, John Haldane draws pointed attention to the fact that evolution presupposes at the earliest stage the existence of capabilities of self-reproduction. Reproduction is an irreducibly purpose-driven act and it has yet to be shown, he notes, that it could arise by natural means from a material base. The British logician Peter Geach comments that there would be no theory of natural selection if not for complex reproductive processes with the strong impression they convey of teleology. And John Maddox makes a telling observation, "The overriding question is when (and then how) sexual reproduction itself evolved. Despite decades of speculation, we do not know."

> *Geek*: I grant that you're not a Vitalist at least in the usual understanding of the term. But you're violating another great taboo of modern biology, and that's importing teleology. If there's one thing that Darwin banished forever from biology, it's teleology. Your argument seems to misconceive the idea of natural selection. Its main thesis is that any organ or capability advantageous for survival will find evolutionary favor. There's nothing more advantageous for survival than replication. Far from casting doubt on evolutionary theory, the reproductive mechanisms establish it as fact!

Guru: It seems to me that natural selection has started functioning as a god of the gaps. You might say that natural selection adapted various organs to perform certain functions, and you could make a plausible case in given instances. But with regard to replication, we're talking of something so fundamental that we've got to go beyond ritual incantations invoking this deity. Of course reproduction has survival value, but the problem is that this capability for reproduction must exist before the struggle for survival can begin. Let me put it another way: *it's the horse of reproduction that draws the cart of natural selection.* If you put the cart before the horse, you won't get started.

So my question remains: How is it that the first living beings had the powers of replication? How is it that life came with this fundamentally purposive capability pre-installed? Replication is the engine that runs evolution. But who came up with the idea of replication, and who then imprinted material structures with a vast variety of replicational capabilities?

> *Geek*: Are you saying that scientists should now be teleologists?

Guru: As a matter of fact, evolutionary scientists implicitly assume purposive categories. In his recent *Darwin and Design*, the neo-Darwinist Michael Ruse considers this paradox: Darwin supposedly got rid of design from biology but then had to re-introduce it as a key concept. According to Ruse, final causes are indispensable for biologists and he illustrates this from the work of evolutionary theorists. Evolutionists look at selection as a mechanism that brings about perfect adaptation, an optimal state of affairs. The design of an organism is appraised from the standpoint of survival and reproduction. Ruse is very clear that he does not himself accept any argument from design to God, but notes that he is a skeptic and not an atheist. He quotes appreciatively from Charles Raven who said that beauty rejoices and humbles him and is remote from words like random or purposeless. In *What Functions Explain,* Peter McLaughlin argues that models of natural selection make the mistake of looking at biological traits purely from an external standpoint instead of seeing how they serve a purpose internal to the systems in which they subsist. Biological organs are not artifacts that you approach as a maker or user, e.g., asserting that wings exist because they enable the survival of the species; rather, they should be seen in the context of the benefit they bring to the individual organism.

> *Geek:* Well I believe the survival of the fittest makes sense of everything in the biological world. In fact, Darwin and most biologists hold that natural selection is the ultimate explanation of all phenomena in nature, and there is no evidence for a Designer behind the scenes.

Natural Selection as an Explanation

Guru: I think it's important that we understand what we're talking about here. From the time of Darwin to the present day, scientists, including evolutionary biologists like C.H. Waddington, and philosophers have drawn attention to the fact that the idea of the survival of the fittest is simply an empty tautology, a circular argument, a play on words. The statement "all bachelors are unmarried males" doesn't reveal anything new; it simply repeats the same thing in two different ways. The philosopher of science, Sir Karl Popper, held that natural selection was unfalsifiable.

In an insightful commentary, Kenneth Gallagher points out that the charge of tautology applies to both Darwinist and neo-Darwinist versions. Darwin held that natural selection favors organisms with traits that fit their environment. When there is a change of environment, selection ensures that only the fittest organisms survive and this sets the course of evolutionary history. Critics pointed out that this means simply that thanks to natural selection the organisms that survive are the ones that survive—hardly new or useful information. The neo-Darwinists adapted the argument to say that genes that add to an organism's fitness for survival will increase in frequency. This turned out to mean that those with the most descendants survive, a mere tautology as Waddington admitted.

Gallagher admits that the theory of Natural Selection as a whole is not necessarily a tautology. The theory says: (a) there is a struggle for existence; (b) there are random variations among the offspring of organisms; (c) some of the variations confer an advantage in the struggle for survival; (d) these are passed down such that the advantageous property tends to dominate within the species; and (e) finally, those with the most advantageous equipment, the fittest, survive. It's only the last statement that's a tautology. In other words, the theory of Natural Selection is not a tautology but the principle of selection within the theory is tautological. Neo-Darwinists like Ruse have tried to stave off the charge by pointing out that several features of the theory are not tautologies. Gallagher's response is that true as this may be, the fact remains that at least one component principle, "the fittest survive," is a tautology, and it so happens that this is the underlying principle. Stephen Jay Gould admits that the neo-Darwinist version is a tautology, but then offers a restatement that he says escapes the charge: "the fittest organism is the one that is best designed." But on closer examination Gallagher finds that this in turn is reduced to the tautology that the fittest organism is the one best designed to leave offspring.

I have dwelt so long on this principle because "the survival of the fittest" is more often than not used as the argument of last resort. When no other plausible rationale for a phenomenon can be given, materialists simply gesture in the direction of natural selection, understood as "the survival of the fittest." But if this principle, as opposed to other components of the theory, is a tautology that doesn't give any new information, then it can't be used as an argument or an explanation. This is not to say that we can't offer an organism's adaptation to its environment as an explanation for a given phenomenon, or draw a correlation between structure, function and environment. These can be valid arguments but they're not the same thing as invoking "the survival of the fittest" as a *deus ex machina*. Evolutionary biologists also acknowledge that there are factors other than natural selection at work such as phylogenetic inertia and developmental constraints.

Geek: But you've admitted that the key driver of evolution, the idea of random variation, is not a tautology. From genetics, we know that the primary vehicle of evolutionary growth is random mutation at the molecular genetic level. Pretty much the entire community of biologists believes that this serves as a sufficient explanation for the whole history of life.

Random Mutation as a Mechanism

Guru: It's true that the existence of a mechanism that generates novelty is fundamental to the success of evolutionary theory. The central candidate for this role in dominant versions of the theory is random mutation, DNA copying errors during genetic reproduction that result in changes to the DNA and genetic code of progeny. The novelty is generated by mutation; natural selection simply enables the survival of the most beneficial mutations. Each mutation happens entirely by accident; there's no purpose, objective or goal driving it and there's no program that commands a specific sequence of mutations to produce a new organ or species. Moreover these random mutations have to take place in the reproductive cells, the gametes, of the male or the female for the formation of mutated progeny with inheritable variations.

Certainly the idea of novelty through random mutation is compatible with belief in a divine impetus behind the whole process. But, in all candor, I have to say that the notion of random mutation serving as the exclusive agent of creative and beneficial change is simply not credible. After all, bugs and viruses in computer software do generate novelty, but it's destructive and not creative in nature. Likewise, there seems to be no question that all observed instances of such mutation have been destructive for the organism. If there are four mutations for every 100,000 gametes, 99% of these mutations are likely to be harmful. You can certainly believe that for every 99 mutations that are harmful there will be one that generates beneficial novelty and given hundreds of millions of years this could be a viable engine of evolutionary change.

But this belief depends upon a match between these supposedly random mutations and the fossil record. Or to put it another way: can random mutations explain what we know of life both today and in the historical record? Here there are two basic variables: the probability of beneficial mutations and the time available. Now probability lies in the eye of the beholder and for every scientist who says that a certain beneficial and enduring mutation is wildly improbable, there may be ten others who say it's perfectly plausible, although improbable. But the fossil record itself is not dependent on such subjective judgments. And this record indicates that all of the 34 animal body plans appeared over the relatively short time span of five million years. I find it inconceivable that random, unconnected mutations at the genetic level could be responsible for the creation of such varied life-forms at such a rapid pace. It seems to me entirely implausible that all of the great transformations and advancements in the history of life can be attributed entirely to random mutations.

> *Geek*: You seem to like giving with one hand and taking away with the other! 99.99% of scientists hold the mechanisms of natural selection and random mutation to be fact and so you granted that they were possibilities. But then you pulled the rug out from under this scenario with arguments of tautology and improbability. Is it your strategy to substitute natural mechanisms with a supernatural agency? Most scientists think that a guiding hand for evolution is not simply superfluous but superstitious. Your critique of random mutation is simplistic. Scientists have successfully recreated genetic mutations, for instance, in fruit flies; by modifying the fly's DNA they've shown how its legs can sprout from its head or it can be born without wings. Computer simulations have also shown the creative role of pure chance. To give one example, Richard Dawkins has shown how computer programs can randomly generate symmetrical life-like shapes called biomorphs that mutate into new species from dots and lines.

Guru: I'm familiar with the visceral fury with which many evolutionists tend to lash out at anyone who raises the slightest doubt about their belief-system. It's almost as if there's an element of religious fanaticism, so that a mere query is perceived as a dangerous threat to the faith. For my part, I have no theological problem with random mutation as the vehicle of evolution; it still leaves open the need for a source of nature and its laws. Similarly I have no problem with the existence of unicorns. I just find both beliefs to be equally fanciful.

The idea of randomness in the progression of life is under attack among biologists. Robert Williams of Oxford and J.F. DaSilva of Lisbon, for instance, have shown that the earth's chemistry is responsible for the path taken by life:

1. The simple structure of prokaryotes, which they retain even today, was a result of the reducing environment in which they were formed.

2. Prokaryotes discharged oxygen by extracting hydrogen from water in primordial times. This resulted in a more oxidizing environment and the subsequent formation of organisms with a nucleus and organelles, eukaryotes.

3. Continued interactions with the environment created the conditions required for multi-cellular structures.

Harold Morowitz, one of the best-known authorities on the thermodynamics of living systems, notes in *New Scientist* that the ideas of Williams and DaSilva are "part of a quiet paradigm revolution going on in biology, in which the radical randomness of Darwinism is being replaced by a much more scientific law-regulated emergence of life."

In a paper titled "Traditional Wisdom and Recently Acquired Knowledge in Biological Evolution," Werner Arber, whom I have cited, notes that "spontaneous genetic variation is generally not directed and often leads to unfavorable mutations that may go along with a serious hindrance in the life of the concerned individual." A fusion of molecular genetics and evolutionary biology, he writes, "implies in particular a departure from the idea that genetic variation largely depends on errors, accidents, illegitimate interactions and selfishness of genetic elements. In contrast, evolution of life has to be seen as resulting from intrinsic properties, forces and strengths of nature."

In reflecting on random mutation and the generation of novelty in the biological world, I would like to recommend to you Gerald Schroeder's *The Science of God* and *The Hidden Face of God,* two superb books by a scientist addressing the state of the field in this area. Schroeder observes that evolutionists now admit that the classic idea of randomness at the molecular level of DNA will not work as the agent of mutation since there simply isn't enough time. He points out that:

1. The body plan of every animal alive today was developed in the five million years between the pre-Cambrian and Cambrian eras.

2. There's no evidence of evolution in the five million years of the Cambrian explosion because every animal of the time emerges fully formed.

In an editorial titled "Engines of Evolution," Peter Brown, editor-in-chief of *Natural History*, asks, "But what about the mechanisms of evolutionary change? What gives rise to the changes that individuals present for testing by natural selection? For much of the history of Darwinism the answers have been genetic mixing through sexual reproduction and random genetic mutation. But the known rates of random mutation have seemed inconsistent with the time available for the observed biodiversity on earth to have evolved."

John Maddox, as fierce a Darwinist as any, comments that virtually nothing is known of the following:

+ how genes switch on the suite of genes hierarchically beneath them

+ how genes give cells their particular nature

+ how the complex system of genes came to be: did it come to be at the Cambrian Explosion or before?

The example of the fruit fly is hardly relevant because it's actually harmful for the subject organism and would leave it helpless against the forces of natural selection. In fact, investigators have yet to find a genetic mutation that results in a beneficial change of body plan. And Niles Eldredge in his *The Pattern of Evolution* makes the startling observation that to date no utterly convincing instance of true speciation involving sexually reproducing organisms has emerged from the genetic lab.

We've already seen why computer games are irrelevant to our discussion. These are at worst exercises in self-deception and at best arguments in favor of intelligent intervention. First, we're dealing here with silicon and not with carbon, with cyberspace and not the real world. Second, it's Dawkins who does the breeding and therefore the selection and mutation; the processes are neither natural nor random. Finally, the exercise demonstrates the need for an intelligent system with no room for error, hardware that works, software that is bug-free and capable of taking instructions.

Dawkins works the same sleight of hand when he tries to show a quote from Shakespeare emerging by chance out of 28 letters after only 43 attempts. In this instance, (a) he includes only the letters that belong to the quote and (b) any time a letter comes to a position in the line that corresponds to its position in the quote he takes it out of the running. This is little better than a game of musical chairs where a winner inevitably emerges as each chair is withdrawn. And far from being random, it's a purpose-driven, rule-based process. All Dawkins shows in his two examples is that you can win any game if you not only set the rules but keep changing them to ensure your victory.

Geek: You've made some good points. But I think you've caricatured Richard Dawkins. As a matter of fact, his metaphor of the selfish gene deals a mortal blow to all delusions of a divine watchmaker. Like Darwin, Dawkins shows that the apparently purposeful design we see in living creatures is entirely a product of natural selection; useful traits enable survival. But unlike Darwin, who focused on the organism as the agent of change, Dawkins zeroes in on genes, while holding that his approach of the differential survival of self-replicating genes in the gene pool is inherent in Darwinism. Genes in living beings ensure their own survival by facilitating a life span for their hosts that is long enough for reproduction. These are the so-called selfish genes, and the diversity of life in the world is a direct result of their ruthlessly selfish struggle for survival.

Fundamentally, as Dawkins puts it, all processes involving life are centered on maximizing the survival of the DNA that makes up genes. Humans are "temporary survival machines." The genes in cheetahs want to improve their chances of survival by giving the cheetah the traits required to chase gazelles, whereas the genes in gazelles want to ensure their survival by enabling gazelles to outrun cheetahs. The elaborate mating rituals in various species are by-products of the attempts of genes to be passed down to successive generations. This can be achieved only if its genes make the male attractive enough to the female of the species. So all activity and all apparent purpose and altruism in nature is driven by the survival of DNA and the replication of DNA sequences. Occasionally genes may cooperate but this is fortuitous, not primary. DNA is indifferent to the suffering of its hosts. It is blind and neither knows nor cares. Any purpose we see in nature is simply a projection of our own habits of thought. At bottom the universe has no design or purpose, good or evil. It's millions of blind selfish genes that explain all apparent purpose in the evolutionary process.

Now Dawkins doesn't believe that we are entirely under the tyranny of genes. He has developed the concept of something called "memes," which are simply ideas formed by humans that can be passed on to future generations. These are cultural—not genetic—units of reproduction and can also enable self-preservation. Dawkins sees religion as the enemy of truth with no evidence at all in favor of it. In fact he calls it a disease of the mind that has the epidemiology of a virus. It's a parasite that lodges itself in the brain and then replicates itself.

Guru: It's appropriate that you conclude with Dawkins' views on religion, which are almost as well known as his theory of genes as agents of evolution! I say this because the picture he's painted of the selfish gene has attracted what can only be called a religious revival in certain circles. Followers fixated by this view of the world have neither considered the evidence in its favor nor discriminated between fact and speculation.

Numerous scientists and philosophers have refuted Dawkins' theory, but none have done it as comprehensively and definitively as Stephen Jay Gould in his *magnum opus, The Structure of Evolutionary Theory.* As Gould sees it, this whole research program is based on a conceptual error that has inspired a quasi-religious following. He notes that the theory faces strong opposition from many evolutionists who see it as a kind of Darwinian fundamentalism. Moreover the theory is fraught with internal errors recognized by its proponents.

The central fallacy of the selfish gene school is the misguided notion that genes are both units of selection and causal agents. They are neither, although they are certainly replicators. It's a fundamental fact of nature, writes Gould, that it's organisms and not genes that live, die and reproduce in the interaction with nature called natural selection. If the organism struggles and survives, then its reward is the greater representation of its genes in succeeding generations. Genes are products, not agents.

Gould lists several arguments that refute the selfish gene theory. The primary problem with the theory is the attitude of mind at its root, reductionism, the wish to explain large-scale events in terms of the smallest players. The mistake is to assume that a fundamental constituent is a fundamental agent.

The selfish gene theory simply won't fly on other grounds as well. Dawkins admits that genes do not have cognition or intelligence and therefore the phrase "selfish gene" is a poor metaphor. Genes are not selfish or unselfish. They are simply bits of code that carry the blueprint for the various parts of our body plans. They have to work together to bring about what's good for the organism. Dawkins might call them "immortals" because of their replication, but in actual fact each individual gene dies in the cell to which it belongs.

Interestingly, biologists tend to be much more skeptical about ideas of purpose and plan, blueprint and code, than cosmologists and physicists. Could this be the case because their purviews of research influence their perspectives? The cosmologists and theoretical physicists focus on the big picture while those who study the manifestations of life on earth have a necessarily more restricted frame of reference.

2. Hominids, Gorillas and Chimpanzees

> *Geek*: I have to agree that you've raised inescapable issues on the mechanisms of evolutionary change. Let me take a different tack: your whole thesis of supernal interventions faces rough sailing when we get to *Homo sapiens*. As you no doubt know, chimps and humans evolved from a common ancestor. *Chimpanzees and human beings, in fact, share 98% of the same DNA*. So the culmination of the evolution of life, the grand finale, is a glorified ape!

Guru: Let me first address the widely publicized 98% issue.

What conclusions do you draw from the fact that we share 98% of our DNA with chimpanzees? In his acclaimed *What It Means to Be 98% Chimpanzee: Apes, People and Their Genes*, the anthropologist Christopher Marks has given us a superb overview of the illicit premises and inferences that lead in general to false genetically-based explanations. The 98% statistic is just one more such red herring. "In the 1990s," writes Marks, "we routinely heard that we are just 1 or 2% different from chimpanzees genetically, and therefore...what? Should we accord chimpanzees human rights, as some activists have suggested? Should we acknowledge and accept as natural the promiscuity and genocidal violence that lurks just beneath the veneer of humanity and occasionally surfaces, as some biologists have implied? Or should we perhaps all simply go naked and sleep in trees as the chimpanzees do? None of these suggestions, of course, necessarily follows from the genetic similarity of humans to apes, although the first two have been proposed within the academic community and promoted in the popular media over the past few years."

So what, he asks, "does the genetic similarity of apes to men mean? What is it based on? Does it have profound implications for our understanding of human nature? Here we will see that the universe of genetic similarities is quite different from our preconceptions of what similarities mean. For example, the very structure of DNA compels it to be no more than 75% different, no matter how diverse the species being compared are. Yet the fact that our DNA is more than 25% similar to a dandelion's does not imply that we are over one-quarter dandelion—even if the latter were a sensible statement. This will be the primary illustration of the confrontation between scientific data and folk knowledge, and of the exploitation of the latter by the former. The extent to which our DNA resembles an ape's predicts nothing about our general similarity to apes, much less about any moral or physical consequences arriving from it."

We cannot draw any conclusions about the relation of anything that chimpanzees do to anything done by humans, says Marks, because we still lack the detailed physiological and genetic data and analyses required. "Since they have been different species for several million years, anything that chimpanzees do may be either (1) an element shared with human nature; or (2) an ancient element of human nature now lost by humans; or (3) an evolved element of chimpanzee nature, never possessed by human ancestors." In brief, as Marks said in an interview, *the fact that an animal shares 98% of its DNA with humans does not mean it is 98 percent human.*

Moreover, most of the DNA we share with chimpanzees is the non-coding or "junk" DNA, DNA that doesn't code for proteins.

Elaine Morgan draws our attention to the big picture: "Considering the very close genetic relationship that has been established by comparison of biochemical properties of blood proteins, protein structure and DNA and immunological responses, the differences between a man and a chimpanzee are more astonishing than the resemblances. They include structure differences in the skeleton, the muscles, the skin, and the brain; differences in posture associated with a unique method of locomotion; differences in social organization; and finally the acquisition of speech and tool-using, together with the dramatic increase in intellectual ability which has led scientists to name their own species *Homo sapiens*—wise man. During the period when these remarkable evolutionary changes were taking place, other closely related ape-like species changed only very slowly, and with far less remarkable results. It is hard to resist the conclusion that something must have happened to the ancestors of *Homo sapiens* which did not happen to the ancestors of gorillas and chimpanzees."

Now let me return to what I said about the history of life. There are three major leaps in the history of life: the origin of life, the origin of consciousness and the origin of self-conscious rationality. The last of these concern *Homo sapiens*. As far as I know, we're the only terrestrial beings who have self-conscious rationality and it's this dimension that differentiates us from all other animals. When did we become self-conscious beings and how? Who knows? Certainly not the paleoanthropologists who have done a fine job foraging through the ancient skeletons in our genealogical closet.

According to current accounts, and these are in a state of constant change, a hypothetical apelike being is the common ancestor of hominids, gorillas and chimpanzees. *Homo sapiens* is the only extant hominid. How many hominids there were in history and how they are related is a matter of controversy. The earliest hominid is thought to be *Sahelanthropus tchadensis*, a species that existed six to seven million years ago, and this verdict may change even as we write. Other hominid species include *Orrorin tugenensis*, *Ardipithecus ramidus*, *Australopithecus afarensis* (of which the famous Lucy is a member), *Homo rudolfensis*, *Homo habilis*, *Homo erectus*, *robustus*, and *Homo neanderthalensis*. Some paleoanthropologists hold that *Sahelanthropus*, *Orrorin*, and *Ardipithecus* belong to the gorilla, chimp and human branches respectively, while some place them all in the same lineage as humans, and yet others classify hominid fossils under even more taxa. The old idea of a steady succession from ape to *Homo sapiens*, of *Australopithecus* turning into *Homo erectus* and then into *Homo sapiens*, has now been replaced by the notion of numerous hominid species competing with each other. Multiple hominid species are thought to have existed in parallel 3 and 1.5 million years ago. Many theorists have replaced the metaphor of a tree of hominid evolution with the image of a tangled bush bustling with parallel hominid branches, some of which turned out to be evolutionary dead-ends.

All we can know about the hominids has to be inferred from a handful of fossils, fragments of skulls and jaws. These fossils have aptly been called bones of contention and, as one reviewer said, paleoanthropologists make up for a lack of fossils with an excess of fury. Since we're talking of events that took place millions of years ago and we have meager data at best, we can't really know how the hominids emerged, how they're related and what they were like. Of course we can use imagination to fill in the gaps, but then we're indulging in speculation and it's hard to distinguish between the fruitful and the fanciful.

But we do know this. At some point *Homo sapiens*, a species distinct from all other hominids, came into the picture, a being with the ability to see and think, walk and talk, paint and symbolize, hunt and farm. While other hominids may have shared some of these capabilities, we know too little about them to reach definite conclusions; we do, however, have definitive knowledge about ourselves. Whether this happened 150,000 or 200,000 years ago is irrelevant. What matters is that the physical emergence of *Homo sapiens* marked the appearance in history of self-conscious rationality, a new level of intelligence that is inexplicable in terms of the purely physical. And just as molecular biologists can identify the chemical substrate of life but not its essential identity as intelligent agency, so paleoanthropologists might identify the physical antecedents of *Homo sapiens* but cannot be expected to grasp with their tools the new realities of rational thought and the irreducible self. As the Nobel Prize-winning brain scientist Sir John Eccles put it, the conscious self is not in the Darwinian evolutionary process but is a divine creation.

Geek: I don't get the point. *Homo sapiens* may well be a separate species, but it's not qualitatively different from its animal forbears.

3. The Existence of the Mind

Guru: I guess that's really the point at issue and I believe only a re-discovery of our self-conscious rationality will show us exactly how and why we are radically different.

All of the activities of science presuppose we have intelligent minds. All of the most popular interpretations of brain activity presuppose that we do not. So which view is right?

Now the most obvious thing of all is this: we exist in two worlds. The first is the world of matter, of fields and their manifestation as particles, stars and organisms, and the second is the world of the mind, of conscious acts and rational judgments, intentions and decisions. Both worlds are radically different from each other but they constantly interact. We cannot explain either one in terms of the other. It's simply ridiculous to suppose that feelings and cogitations are nothing but photons and electrons. Scientists by their very profession focus on the first and discount the second, although implicitly for all their theories they rely on the reality of rational thought.

In my view, none of the theories of the origin and evolution of the universe or the structure of the microphysical world can explain or touch on the issues of mind and conscious awareness. Unfortunately, many scientists are *unaware of their awareness, unmindful of their minds, unconscious of their consciousness.*

It's my contention that there are four features of our experience as human beings that cannot be understood solely and simply as physical activities. Certainly we use our brains and nervous systems in performing these actions, but by their very nature the capabilities listed below are essentially immaterial:

1. Being conscious, perceiving
2. Thinking, forming concepts
3. Being aware of our identity and self-hood
4. Intending and planning, choosing freely

Humankind as a whole has recognized the trans-physical nature of these properties. So have most philosophers until Descartes. Our experiences of subjectivity, identity, intentionality, insight and thought are not only undeniable but they are undeniably different from all our experience of the physical. Thoughts, intentions and beliefs are in their very essence conscious acts that cannot even be conceived of in material terms. And once we realize the intrinsic immateriality of these acts and experiences, we realize that they could not be products of matter.

Geek: Although you mention matter in passing, you talk as if we're pure spirits, not even spirits imprisoned in brains. Armies of neuroscientists, cognitive psychologists, Artificial Intelligence researchers and anthropologists have established materialism as *the* scientific view. We are animals different in degree but not kind from the great apes. What we call activities of our minds are simply operations of our nervous systems. We're obviously conscious beings, but researchers like Francis Crick have shown the precise physical correlate of consciousness. In fact, I don't think you'll find a single brain scientist who thinks that we have a separate non-physical part located somewhere in the brain barking out commands to the body. As one of my colleagues once said, minds are simply what brains do. They don't exist in different worlds. They're just different ways of describing the same thing.

Guru: This last idea of minds being what brains do is a textbook version of what might be called naïve materialism. Brain scientists tell us what the brain does when we see or think. A chemist who analyses a work of art using microscopic paint analysis and ultra-violet and infra-red examination methodologies can tell me about all the physico-chemical constituents of the painting. I don't expect him or her to have any special insight into the aesthetic content of the painting, its impact on us or the intention of the artist.

The naïve materialist argument reminds me of an unintentionally comical article on mind and body. Its author said that brain researchers who've spent a century looking for the self have concluded that there's no place where it could be located in the brain and therefore there's no such thing. This conclusion is comparable to the remark of the first Soviet cosmonaut to orbit the earth who said God doesn't exist because he couldn't be found in outer space. In both instances, the investigators had not only looked in all the wrong places but had entirely misunderstood what they were looking for. Neither God nor the self have ever been thought of as occupying space, and, as a matter of fact, great brain scientists of the last one hundred years, ranging from C.S. Sherrington to Wilder Penfield to John Eccles, accepted the existence of a trans-physical reality separate from the brain. The very idea that the self, if it exists, must be located in the brain is on a par with the notion that we can find time by dismantling a clock.

To be taken seriously a theory must adequately explain the relevant data. This materialism has failed to do because it is plainly just a belief with no generally accepted theory or mechanism to support it. Remember, there's a graveyard of rejected materialist theories, none of which materialists themselves seriously propose; Skinnerian Behaviorism and 19th century Mechanism are prominent examples. The materialists, physicalists, determinists, mechanists, behaviorists and functionalists have scurried from one argument to another as even they eventually disavow each new attempt to deny the obvious and try to come up with yet another argument in favor of the idea that mind is just matter. The very fact that they have to keep inventing new arguments, all of them based on unverifiable speculation, is an argument against the overall soundness of their futile enterprise, since materialism seems to be in a state of perpetual retreat.

Materialism is demonstrably helpless in addressing the issues of perception, self-consciousness, intention, identity and language, and some of our best modern thinkers have developed incisive critiques of materialism from these uncontroversial starting-points. David Lund argues, for instance, that materialists have been unable to explain either what it is to be conscious or to have an experience; once they admit the existence of a subject of consciousness, which they can hardly avoid, then the whole case for materialism is lost. Richard Swinburne notes wistfully that a lot of philosophical ink has been wasted trying to deny what stares us in the face: the distinction between conscious events and brain events.

> *Geek*: Well, it could be said the major breakthroughs in science came when scientists risked ridicule to scientifically explain phenomena that were previously attributed to mysterious supernatural forces. A biological mapping of the mind may not be complete as yet, but it's only a matter of time before we lay out a comprehensive roadmap of all mental activities entirely in terms of neural behavior. Meanwhile, there's no credible explanation of these events as non-physical in nature.

Guru: As you rightly acknowledge, physicalists and materialists still do not have what you call a "mapping of the mind." John Maddox concedes that an understanding of how the brain works is the second of two virtually untouched problems in modern biology, the other being the origin of life. While noting that great progress has been made in cataloging the properties of brain cells and structures, he says we're a long way away from identifying the neural circuits responsible for both simple responses to sensory information and higher functions such as remembering, imagining and choosing.

I commend Professor Crick for taking consciousness seriously, but I think his electrophysiological explanation of consciousness is as flawed as every other reductionist account of experience. Critics have noted that the best a theory like Crick's can do is show that certain mental processes can be correlated with certain brain activities, but such a theory can't and doesn't pretend to show why there's a subjective experience that accompanies the physical operation. Crick admits that he has no answer to the question of what causes the specific experience of being conscious. And his theory as I understand it is concerned simply with visual consciousness, not with higher operations like conceptual thought, self-identity and intention.

In *How the Mind Works*, Steven Pinker, one of today's foremost writers on the brain, makes a frank assessment of the situation with regard to consciousness. We have a lot of data on consciousness in the sense of access to information. We know the categories of information processing carried out by the nervous system and we have some hypotheses, such as Crick's, about what brain mechanisms enable consciousness to access such information. But we have no idea where sentience, i.e., *what consciousness feels like on the inside*, came from. Nor do we know how or why. Pinker says that *neither the computational theory of mind nor any finding in neuroscience have an answer* to these basic questions about our subjective experience of consciousness. He notes that thinkers like Daniel Dennett, who try to deny the reality of consciousness, have not really addressed this core issue in their arguments. In his view these are valid questions but we can't answer them because we don't have the cognitive equipment required.

Saying that we have no scientific explanation of sentience, says Pinker, is not saying that sentience does not exist at all: "I am as certain that I am sentient as I am certain of *anything*." He adds that sentience is not a combination of brain events or computational states. In my view, the only legitimate resolution to the problem of consciousness is an ontological one.

Moving on, the full absurdity and indefensibility of the thesis that there is no such thing as thinking should be self-evident if we just stop to think about it, and thinking, in this context, would mean a clear consideration of the issue on its own terms without attempting to explain away the hard facts of everyday experience.

The data of our daily experience may be summarized thus: we are conscious and aware that we are conscious. We:

- perceive
- conceive
- remember
- imagine
- sense
- feel
- plan
- intend
- choose.

Our experience is not only qualitatively subjective in nature (consciousness) but also experienced as *our* experience (the self) and experienced *by* us (intention); we can describe and discuss things that go beyond anything that we directly experience (intellect). Moreover, we know experience itself as irreducibly different from all that is physical in nature.

Why talk of anything distinct from the brain, you might ask, when a study of its regions and functions shows a close correspondence to most mental events? After all, scientists have done an outstanding job identifying the networks and connections and entities that control memories and sensations and the like. And if you insist on talking of a mind distinct from the brain, how do you visualize it? These are both difficult questions but the only satisfying answers are those that can be reconciled with our experience, since everything we know about the mind comes from experience and everything in experience comes through the mind.

1. The **first** datum of experience is that *brain and mental events have entirely different properties.*

2. The **second** is that *mental events cause brain events.* In all cases we don't frame a conscious intention to perform an action. Sometimes

our intentions are subconscious. But clearly we do act on the basis of conscious intention in most instances.

3. The **third** is that the brain-and-mental-event correlations do not include the "I" who is responsible for both brain and mental events. *There is no location in the brain where the "I" is located.*

The problem of visualizing the mind is hardly a unique one. We can't visualize the energy fields that are so basic to physics or the pre-Big-Bang universe or the time dilation of the Theory of Relativity. So why should we expect the mind that thinks about all these things and applies them in various contexts to be any less difficult to visualize?

We may not be able to visualize the mind but we can certainly describe the central characteristics of our experience as self-conscious rational beings: consciousness, thought, language, the self, free will and intention. This is what I propose to do next.

Consciousness and Thought

We perceive and we conceive; these are the two primary and distinct activities of the mind, one sensory in nature and the other intellective, although the activity of perceiving with the senses often requires intellective analysis as well. The activity of the intellect is especially apparent in the classification of our experience under universal concepts; the mind works on the data of our immediate experience to see what's similar between different objects and to separate these similar elements into universal concepts, e.g. the concept of a dog from our experience of Rover. However, the universals have no separate existence apart from the instances in which they're embodied. Also not all concepts are directly related to sensory data; the concept of God, for instance, explains the existence of the world of sense-data but is not itself a part of sensory experience.

Concepts spell the death knell of materialists since there's no way to understand them in terms of matter. John Haldane, for instance, has shown that concepts transcend material configurations in space-time because (a) any property of a thing can be described in many non-equivalent ways of thinking, (b) this order of concepts is thus far more abstract than the natural order, and (c) it's hard to see how the former then could spring from the latter. Tom Sullivan notes that all human thought revolves around our ability to effortlessly think of universal characteristics that can have local instances, for instance the ideas of hot and blue. This is clearly an immaterial activity because these ideas are not linked to something physical. Think, for instance, of blueness.

Here I might add a comment on comparisons of humans to computers and animals.

Computers process information just like the intellect. But this doesn't mean a computer is the same thing as a human intellect given that the computer isn't conscious or aware of itself doing the processing, as we are when we think. Nor does it have insights and intuitions as we do or understand the meaning of the symbols it processes. All of its activities reduce to electrical impulses; the very idea of self-conscious thought is obviously inapplicable. Finally, the computer only performs the activities it was programmed to do by its human creator.

Animals use sounds as signals for communication and never as signs that refer to concepts, as is the case with most words we use. The higher animals share some of our abilities of sense-perception and possibly even abstraction at this level. But all the abstraction is limited to things they have perceived with their senses. Experiments with chimpanzees, bonobos, monkeys and other animals have shown that they respond to certain physically observable things, perceptual objects, when these are displayed. But there has been no evidence that they can deal with concepts and abstractions unrelated to the sensory realm—this goes for Koko and Kanzi!

None of the so-called linguistic behavior in animals is to be found in nature. Heini Hediger notes, "With all animals with which we try to enter into conversation we do not deal with primary animals but with anthropogenous animals, so-to-speak with artifacts, and we do not know how much of their behavior may still be labeled as animal behavior and how much, through the catalytic effect of man, has been manipulated into the animal." In the case of many of the apes "it is the production of certain signs in which we would like to see a language. But how can we prove that such answers are to be understood as elements of a language, and that they are only reactions to certain orders and expression, in other words simply performances of training."

"Language is, indeed, the ultimate symbolic mental function, and it is virtually impossible to conceive of thought as we know it in its absence," says Ian Tattersall. In his view, "When we speak of 'symbolic processes' in the brain or in the mind, we are referring to our ability to abstract elements of our experience and to represent them with discrete mental symbols. Other species certainly possess consciousness in some sense, but as far as we know, they live in the world simply as it presents itself to them.

"Members of other species often display high levels of intuitive reasoning, reacting to stimuli from the environment in quite complex ways, but only human beings are able arbitrarily to combine and recombine mental symbols and to ask themselves questions such as 'What if?'"

We cannot deny that there are primitive forms of consciousness and communication in animals and marvelous inbuilt instincts. But the point I wish to make here is that they show no trace of conceptual thought; no animal can have the idea of God or the concept of a square root.

Another difference between humans and animals, notes James Miles, is the total absence of true love in the animal kingdom. Darwinists like George Williams point out that cannibalism and infanticide are universal except in vegetarian species. The mother of the chimpanzee that fights fiercely to defend its child will mate with its child's killer. Likewise, says Miles, there is no altruism, compassion, extreme self-sacrifice, or laughter. Animals tamed by humans sometimes appear to acquire certain human-like attributes, but this is by no means natural to them.

Language

We have seen how syntactical language, an activity we indulge in without effort or a second thought, is unique to humans and inexplicable in terms of evolutionary history. I have also already spoken of language as a system of codes that conveys meaning through symbols. The activities of coding and decoding, of seeing meaning, are irreducibly immaterial.

In a splendid study of human language, Herbert McCabe observes that language is built around the ability to understand. There is no organ, no part of the brain that performs understanding. While an animal's behavior is a response to the experience of its senses, the human goes beyond mere responses to sensory experience. The human is able to interpret the experience of the senses and communicate this understanding to other humans. It does so with language, which is an invention involving symbols to which we give meaning. Language is not something that can be genetically transmitted. It is acquired anew by each person through a process of learning (understanding), of course building on what has already been constructed by the community. The atheist philosopher Antony Kenny remarks that language can't be explained by natural selection. The idea that language-users have a selective advantage is meaningless because you can't be a language-user unless there's already a community of language-users. This idea is tantamount to saying that golf evolved because golf-players had an advantage over non-players.

McCabe points out that animals may make noises and gestures for communication, e.g. whales creaking, but these are pieces of behavior that have a function in their life-pattern and are genetically determined. Certainly animals are capable of imitating our activities to some extent, especially if there's a reward for doing so. The psychologist Herbert Terrace, an ape communication research veteran, concludes that animals are not capable of humanlike language; rather, they use sophisticated methods to request things. Their apparent uses of language are driven by physical desires of some kind.

In *Unweaving the Rainbow,* even Richard Dawkins draws attention to the mystery of language. He points out that nobody knows how language began. "There doesn't seem to be anything like syntax in non-human animals and it is hard to imagine evolutionary forerunners of it. Equally obscure is the origin of semantics, of words and their meanings." Amazingly, human language has open-ended semantics that enables an indefinitely large dictionary of words and an open-ended syntax that enables words to be combined in an indefinitely large number of sentences.

"Chomsky," writes Stephen Clark, "who did so much to identify what is needed if anyone is to speak or understand a language, denies that human languages can be derived from animal communication systems, precisely because our languages depend upon syntactical rules." Chomsky introduced the idea of a Universal Grammar to which all humans have access. Says Clark, "It seems overwhelmingly likely that Universal Grammar is a species-specific feature of humanity, just like learning to walk." This doesn't fit into the evolutionary framework because "there is no precedent for Universal Grammar," continues Clark. "It is a single, shattering mutation. Non-linguistic systems of communication allow only a finite number of discrete communications: human languages permit an infinite number. But such rapid shifts are not ones that evolutionary theorists can accommodate: the point of neo-Darwinian theory is that change from one generation to the next is gradual, that there [are] no really novel organs or behavioral patterns despite the differences eventually made."

The Self

We have also already highlighted the three dimensions of the self:

1. It is the center of our consciousness and unifier of our experiences

2. It maintains our continued self-identity over the course of our lives (individual brain cells are just as incapable as the chips in a computer to serve as a self)

3. It is the basis on which we act as self-conscious free agents.

In *Infinite Minds,* John Leslie points out that a computer that exactly duplicates the information processing found in the brain would still not be a conscious self. Its components are obviously distributed over different locations and do not and cannot have the particular kind of unity characteristic of our conscious experience.

I've pointed out also that there's no part of the brain where the self is located. Now some materialists have said that there'is no royal I, just a number of agents in the brain that direct different activities. Steven Pinker retorts that this doesn't make sense since we have only one body and multiple agents would make it impossible for us to function at all, since there's no unifier. Certain parts of the brain, for instance, the frontal lobes, may be essential in decision-making processes but, as Pinker says succinctly, "The 'I' is not a combination of body parts or brain states or bits of information, but a unity of selfness over time, a single locus that is nowhere in particular."

Free Will

That we think we make free choices all the time is undeniable, whether it's a question of eating out this evening or of agreeing to make a donation. The only question is whether these choices are truly free, as we believe them to be. If materialism is true, if matter is all that exists, then there's no free will. But our experience of free will, on its own, is a sufficient disproof of materialism if we attend to it on its own terms. We know that we're faced with alternate courses of action all the time and we know that we choose between these alternates, sometimes after deliberation and sometimes with very little or no thought. The determinist tells us that even those choices we made after much deliberation were not made on the basis of our weighing of reasons pro and con, but were the inevitable results of physical states of affairs going back to the very origin of the universe. But this explanation, it has often been pointed out, cuts both ways: if there are no reasons that ground any choice or decision, then there are no reasons that ground determinism. Thus the argument for determinism, like the argument for materialism as a whole, collapses into a metaphysical mud-slinging match: physical causes on the one side and reasons on the other.

Obviously computers process information and perform logical operations, and there are only causes at work here, not reasons. But, to repeat what I've said, the computer only does what it's programmed to do. And the computer itself cannot see the meaning of its operations or weigh the reasons and come to a conclusion. All that's happening in a computer is a series of electrical operations. It's only the human who can make sense of what results from these operations.

The fact of the matter is that our actions can quite clearly be explained in terms of reasons and purposes, not simply in terms of our brain states. Heredity and environment certainly have a part to play in the kind of decisions we make, but these are not the whole story in all of our decisions. There are choices and decisions we make after weighing attractive alternatives, and we make right and wrong decisions sometimes against our wills. Some philosophers turn to determinism, the view that physical processes predetermine all events, because they believe that the only other alternative is to attribute all actions in the world to purely arbitrary factors or to chance. In other words, you have to believe that the actions are uncaused. But the choice isn't simply between physical predetermination and total arbitrariness. There's a third option: reason, the exercise of the will under the influence of the intellect (of course there is a use and an abuse of reason). Human beings cause their own actions—their actions are not simply arbitrary—but their causing of these actions does not originate in a physical event.

The old determinism assumed that all physical events in the universe were in principle predictable. Quantum physics has shown us that at least at the subatomic level there can be no exact predictions—only statistical ones. But this doesn't of itself show the reality of free will. The kind of reason-driven actions we find in free will is different from the probabilistic predictions of quantum theory. Likewise, Gödel's Theorem, especially as expounded by J.R. Lucas, has shown the inadequacy of the Mechanist model when applied to the human mind and to free will. But the ultimate basis of free will is our experience of being agents who make up our minds on various matters and make decisions for which we are responsible.

The idea of responsibility is especially relevant here. If we have no free will then we're not responsible for any of our actions. Murderers and embezzlers can't be punished and heroes and altruists can't be praised because all of their actions were caused by purely physical factors over which they had no control. The whole fabric of society would collapse if we acted as if determinism were true, not to speak of the trivialization of such moral outrages as the Holocaust. This doesn't prove the truth of free will, but it certainly indicates that determinism can only be preached, not practiced.

Only in recognizing a trans-physical center of all our willing can we make sense of our experience, and also of the basis of our responsibility. All acts of the will are trans-physical in nature, although some of them require the cooperation of our bodies; these acts are not uncaused but have reasons and purposes rather than physical processes as their driving force, and all acts of the will are acts of a person. Free will doesn't introduce arbitrariness and uncaused events into nature. Free will instead shows that there are mental causes for mental events and that, as we already know, we are truly responsible for our choices.

Purpose and Intention

Closely related to free will is the reality of intention, another trans-physical activity, in our lives. Now purposive behavior, which is lethal for full-blooded materialism, is evident at all levels of life, and not just human life. The manifestation of purpose in the world reaches its summit in the intentional behavior of human persons. Every person is aware of pursuing goals, of intending to act in certain ways and then carrying out the intention. Such intention cannot be a purely physical process since the very act of intending something implies a deliberation that is intellectual and a deliberation that is performed by a self. Intention is not restricted to abstract planning sessions but covers the most trivial everyday activities, for instance, intending to bend down and tie one's shoelaces. But, in observing intentional acts that require thought and deliberation, we recognize that these acts are trans-physical in nature since they involve acts of the intellect, conceptual thinking as opposed to simple responses to sensory experience, acts indeed of a personal consciousness. Peter Geach points out that words like "to say" are inescapably intentional, something which the materialist cannot accept. As with the denial of other irreducibly immaterial phenomena, any attempt to deny purpose and intention has a boomerang effect. Alfred North Whitehead once said that it would be interesting to study those who devote themselves to the purpose of proving that there is no purpose.

The Origin of Mind

If the evidence for the existence of the mind is so fundamental in our experience, why are so many scientists materialists? According to Sir John Eccles and Daniel Robinson, it was Darwin who misled succeeding generations of scientists with his naïve assertion that thought is a secretion of the brain much like gravity is a property of matter. Although no scientist now accepts this view of thinking, materialists have kept inventing arguments to defend the unproven idea that our mental experience is material in nature. These new arguments, based as they were on unverifiable speculation, have eventually died at the hands of everyday experience. C.D. Broad once said that Behaviorism, a variant of materialism, belongs to a class of theories that are so "preposterously silly" that only "very learned men" could have thought of them. The theories gain acceptance, he went on to say, because they are presented in highly technical terms by learned persons who are themselves "too confused to know exactly what they mean."

Today's materialists no longer deny that we do experience consciousness; they simply try to explain it away as the inner side of outer brain events. This ploy simply begs the question, because we want to know how a set of physical events can result in a very real non-physical experience. The materialists have to defend the self-contradictory conclusion that a single entity can at one and the same time be truly mental and truly material. Even worse for them, it's clear to all that the non-physical experience, the thought, comes first and thereby causes certain physical events. It's not neural firings that cause a thought; the thought drags the neural cart behind it. But where did it come from and how? What causes the mental events that cause the brain events? Other brain events? Then, is the thought "our solar system has nine planets" simply a set of brain events caused entirely by other physical transactions in the brain? Clearly it's the mind acting on the brain that's the source of all our experience.

Nevertheless, the materialists' acceptance of consciousness as a datum marks progress. But it presents them with a dilemma: how can a purely physical universe give rise at random to something that we experience intentionally as qualitatively non-physical? We have said that it's simply incoherent to assert that a universe of pure matter, with no purpose, no intellect, no consciousness, no will whatsoever, can give rise to conscious, thinking, willing agents. Once we admit that we're conscious of being conscious, then we're getting closer to a recognition of the inevitability of an infinite Intelligence that grounds all mental experience. Leading participants in contemporary discussions of consciousness, e.g., Colin McGinn and David Chalmers, have said that it's difficult to see how consciousness could rise spontaneously from insentient matter.

Geek: Which brings us to the question of your model of origins. You've identified problem areas in standard biological models of genesis and novelty. Creationists tend to snipe at evolutionary theory without offering a plausible alternative of their own. You've promised to present your own picture of how things originated. So stand and deliver!

Guru: How did life, consciousness and mind come to be? I'm going to jump the gun a bit by turning now to this question, although it's really a part of my discussion of the laws of nature.

❧

Chapter

Interacting with Infinite Intelligence

S I SEE IT: every quantum leap in the IQ of the Universe, first the coming of life, then consciousness and finally self-conscious rationality, is introduced by an input commensurate with the output. And every such case involves an increase of intelligence: life in contrast to non-life is intrinsically intelligent because it centers on purpose-driven, dynamic agents; consciousness is a higher level of intelligence because it engenders awareness of the agent's environment, a primitive form of knowledge; and self-consciousness and rationality are obviously the highest forms of intelligence encountered in our own experience.

Now there are three ways in which we could explain the appearance of these three new realities. First, we could deny they exist by saying they're all simply different ways in which matter manifests itself in certain conditions and environments. Second, we could say that matter has innate self-organizing properties that generate increasingly complex systems from very simple components. Third, we could draw the obvious conclusion that the source of any increase in intelligence must at least be intelligent.

As you know by now, I find the first two options unbelievable and therefore unacceptable. There are obvious gradations of intelligence in the hierarchy of life, from unicellular to conscious to self-conscious, rational life.

The philosopher of science Sir Karl Popper noted that the emergence of consciousness in the animal kingdom is perhaps as great a mystery as the origin of life. For my part, I can't see how the incredible intelligence of the most primitive forms of life, e.g. their reality as agents, their mapping of complex symbols in performing even mundane activities, could emerge from inert matter. Still less can I comprehend the view that there's no essential difference between life and non-life, consciousness and the lack thereof. I prefer to take the body of data as it stands, and not how I'd like it to be. And the data tell me at least this much: intelligence exists and its existence can only be explained by an intelligent source.

So my starting point is that every increase in intelligence is inexplicable without an intelligent source. Which leads to the obvious question of how this intelligent source manifests itself, specifically what's the mechanism it uses to maximize IQ.

The best analogy I can think of in this context is the relationship of hardware and software. Computer hardware itself is simply an aggregate of silicon. Software progressively transforms these mindless boxes into information processors performing calculations and other complex operations. The hardware responds to electrical inputs alone. Its circuits are either on (1) or off (0). All the data stored on the computer and all its operations ultimately have to translate into combinations of these binary digits. This string of 0s and 1s is called machine language, the lowest kind of computer language. Higher-level software programs convey complex instructions that are translated into machine language and then executed by the computer. These programs are of two kinds: systems software that controls the operations of the computer, e.g., its operating system, and applications software that is used to perform specialized functions. An example of systems software is the operating system that launches and monitors programs and processes inputs and outputs. An example of applications software is a word processor used to write a document or a spreadsheet used to process numerical data.

As I see it, matter can be seen as the hardware platform, life as the operating system and consciousness and rationality as complex applications running on the system. Each software program contains coded instructions that tell the computer to perform certain operations and provide new informational outputs. If life is the program that monitors the hardware and keeps it running, consciousness elevates the system to a new level of information processing and rationality takes it still further to logical operations and abstract thought.

The computer with or without any software program has the same hardware configuration. A computer running the most sophisticated applications can be viewed simply as a box. Thus any living thing can be seen as just an aggregate of chemicals or matter fields. Its intelligence is revealed in the operations and tasks it is capable of performing.

Now the hardware platform can't bring software into being and hardware can't run without software. New layers of software can be added to a given piece of hardware to generate new functions and greater levels of complexity. A PC that runs on the Windows operating system can perform new tasks as different kinds of application software are loaded, e.g. word processors, spreadsheet programs and the like. *The progressive manifestation of intelligence, life and self-conscious being in the universe can be best understood within this framework of new software programs being loaded on the same hardware platform.*

At various points in the history of life, there have been quantum leaps of intelligence. Each such leap is a leap of kind, not degree, as is evident in the various kinds of intelligence perceptible by us in our immediate experience. There's a clear progression in complexity in the organization of the things of the world. In parallel, there are fundamental transitions where new ecosystems of intelligence and energy come into being. The old and the new ecosystems are so different that the first cannot explain the second. Biological history is a drama of continuous overlays of higher levels of intelligence, of new instruction-sets introduced in the matrix of being. What is responsible for this progression? Mass/energy or intelligence? It seems clear that intelligence is the fundamental reality and mass/energy is the lowest rung of the ladder by which intelligence introduces itself.

In my view this is the only model that makes sense of all the data because it acknowledges the hierarchies of life and intelligence. This model incorporates the insights of modern cosmology and molecular biology while accepting under its umbrella the undeniable supra-scientific realities of existent being and intelligence, life and consciousness. The model is a product of the framework of thought I've called the Matrix. It incorporates the synthetic paradigms of great scientists like Einstein, Planck and Heisenberg and thinkers like Avicenna, Maimonides, Aquinas and Madhvacharya who were driven by a vision of intelligence and rationality underlying the universe. Darwin himself at one stage held that "laws impressed on matter by the Creator" drove evolution. Neo-Darwinism, at any rate, is concerned with the origin of species, not the origin of being or of the laws of nature for that matter. You can keep mutating indefinitely, but what you mutate *into* must already be present as a potentiality from the start. The argument from random mutations does not therefore address the question of origin: how did these potentialities and the laws applying to them get there?

The punctuated equilibrium hypothesis of Gould and Eldredge works well with the software/hardware model. Although Eldredge and Gould themselves saw no role at all for any transcendent ground, their model shows that (a) there is a pattern of sudden discontinuity in the history of life, (b) these discontinuities produced life as we know it, and (c) the discontinuities cannot be explained in neo-Darwinist terms and are so radical in nature as to require new perspectives.

Even some prominent neo-Darwinists recognize the reality of radical transitions and the progressive increase in intelligence. In *The Major Transitions in Evolution* and *The Origins of Life,* a leading evolutionary theorist, John Maynard Smith, highlights seven such transitions: the origins of metabolism, self-replication, the genetic code, cells, eukaryotic cells, sex, multicellular organisms, animal societies and language. As he sees it, it's all a matter of increasing complexity in the way in which information is stored, transmitted and translated with biological replicators playing a key role. In a June 2003 interview with *New Scientist,* Smith (who is an atheist) even says that "the analogy with a computer program seems to be a good one." He notes that "it is a theory of information—writing instructions for making things happen."

> *Geek:* If your point is that infinite Intelligence lies behind everything, you must give a concrete model of how this intelligence brought things about. Are we talking creation in six days, one species at a time? Also are life and mind substances that exist separate from matter?

Guru: Remember that the standard evolutionary model postulates a sequence of events in which things advance and ascend up the scale of intelligence through sudden bursts and jumps. The difference between the standard position and mine is simply that the former attributes these night-and-day transitions to random and wholly accidental transformations mysteriously undergone by lifeless, mindless matter, while I persist in calling a spade a spade. If something shows signs of intelligence, we should start with the assumption that it proceeds from an intelligent source. Only a prior ideological commitment can lead us to automatically assume that life came from non-life or mind from non-mind when all of our everyday experience tells a different story.

I hold that every manifestation of intelligence is ultimately grounded in infinite Intelligence. Does this mean that if we were present on the primitive earth, we would witness the divine Hand gradually fashion new forms of life and intelligence or a cosmic watchmaker construct various kinds of clocks? By no means. It's this anthropomorphic conception of God that has been the target of the atheists, and sadly it's the idea of God that many creationists espouse. The actions of infinite Intelligence involve change only in creatures. When a teacher presents some great truth and a listener suddenly grasps its implications, the change in knowledge takes place only in the student, not in the teacher. There's no succession of processes in the divine Mind concurrent with the transitions and leaps in the world; the changes take place entirely in the world, driven though these are by a transcendent blueprint.

Of course, all matter, life and intelligence proceed from the Godhead. But how and when? I know you will say that talk of divine creativity is meaningless if there is no tangible set of actions involved. Science, you will say, has the merit of testing and measuring the quantitative and can therefore confidently predict verifiable results. But as you well know, science operates within the parameters of the laws of nature. It can't tell us how these laws originated or when they kicked in. Science needs intelligence in nature to work, but science can't tell us how this intelligence came to be. Was there ever a time when these laws didn't operate? What would you see when the laws were introduced? What scientific procedure can observe and measure the introduction of the laws of nature, and if these are not themselves measurable are they therefore meaningless?

The key to the whole creative process is the progressive introduction of new laws that organize and generate new realities. Just as the laws of the universe, the intelligence of inanimate nature, came to be at the same time as the universe, so also the intelligence that is life, consciousness and rationality came to be simultaneously with the emergence of living organisms, animals and humans. All three kinds of intelligence are by their very nature intangible and so their introduction in the world was necessarily intangible. Nevertheless, all three are real and have concrete effects on the world around us.

You ask if life and mind are things that are different from and additional to matter.

About life let me say this. There's no thing called life; it's an abstraction like redness. There are only living beings. Living beings are autonomous agents that process information, generate energy and replicate themselves using the incredibly intelligent symbol-processing system we call DNA. When a bug dies its material components remain in place but something's no longer there, something that constituted its existence as a bug. It would be too simplistic to call this its life-principle. It's the entire reality of being an agent and functioning as an intelligent system that's missing. This is a reality that uses a configuration of matter but isn't merely the configuration, since the configuration remains after the bug dies. Now how did that reality come to be? Of course, in the case of any particular bug, it's obvious that its origin can be traced to two other living agents. But when you take the chain of life back to the beginning, you ask how the first *living being* came to be. To answer this you have to address what it is. Once we see life as a reality of autonomous agency and intelligent processing, we realize that such a reality can only come to be if it comes from a source that's not just intelligent but also an agent.

In the case of mind, I've already noted that it's fundamentally distinct from matter in its operations. Mental operations involve neural transactions, but it's clear that there's no coherent way in which we can identify the two with each other. Thoughts and insights have no physical properties, although they can't take place without correlated physical activities. And it's not the material structures that drive the thoughts; rather, our thoughts cause the changes in the brain. The thinker of the thought, the thinker that decides what material transactions are to be carried out by the brain, itself clearly can't be a purely material entity. Here we don't have to ask whether the chicken or the egg comes from first. It's the thought that comes first. And the agent capable of abstract thought and ingenious insight, idealistic sacrifice and unconditional love, is a reality that's utterly different from both matter and all other living beings. Of this reality we can say with conviction that its existence proceeds directly from infinite Intelligence. Haldane, who speaks of a "first cause of conceptuality," notes that there are no intermediate members between "human conceptuality and its causative source, the mind of God."

The materialist idea that consciousness and mind are simply computational processes leads to conclusions its authors may not have anticipated. "There is a more fundamental problem in using today's software methods as a paradigm for the emergence of human sentience," writes Ellen Ullman. "Software presupposes the existence of a designing mind, whereas the scientific view is that human intelligence arose, through evolution, without a conscious plan. A 'little man,' a homunculus, lives inside software. To write code, even using 'object oriented methods' that seek to work from the bottom up, someone must have an overall conception of what is going on. At some level there is an overriding theory, a plan, a predisposition, a container, a goal. To use the computer as a model, then—to believe that life arises like the workings of a well-programmed computer—is to posit, somewhere, the existence of a god."

> *Geek:* But how exactly does infinite Intelligence act in this model? If its action is too abstract to be described, then how does this model differ in any significant sense from the view that all things originated through a combination of chance and necessity, i.e., without any intervention from the outside?

Guru: As I'm sure you'll appreciate, these are matters that can't be reduced to sound bites. Again let's review what atheists and theists agree on. In the history of the world there is a progressive manifestation of increasing intelligence and complexity. Along this path, certain events represent advances in intelligence. We agree such events took place but we disagree on the cause and mechanism and nature of these events. Here I want to emphasize that when there's a fundamental transition, a radical change, then any explanation of this transition must match the facts at hand if it's to be plausible. The burden of proof is on the person who denies the obvious, and here the obvious is the framework of intelligence.

In essence, the theist position is that infinite Intelligence manifests itself in the transition events and in the laws underlying the whole creative process.

To the question of how precisely the supreme Intelligence intervenes in this process, I would suggest that you examine how intelligence manifests itself in our experience. The first step is to become aware of intelligence present everywhere, to reflect on what it is that grounds this intelligence, and then to realize that this ground, infinite Intelligence, is active HERE and NOW. It's active now with the same force and power and creativity as at all the major transitions in the history of life.

Now we cannot look to biologists and other scientists to give us the paradigms and models required to understand this progressive manifestation of intelligence since this is a meta-scientific enterprise. The scientists work on the hardware but here we have a software issue. True, some of the hardware specialists deny there is any software to speak of, but there's no reason why we should take them seriously. With purely scientific tools and terms, we can't understand the nature and origin of purposive structures like reproduction, conceptual structures like language and ontological structures like laws of nature, intelligent agents and self-conscious beings. These are realities that have to be analyzed, understood and explained on their own terms.

The crux of the atheist mistake may be illustrated with our hardware/software model. When you look at a computer doing a task, you see only hardware, not the software or even progressive layers of software. We may not even agree that there is software. But the facts are:

1. Hardware and software are qualitatively different.

2. The software is present here and now.

3. The software was introduced in different stages.

The application of this analogy is simple: the material world is different from life, consciousness and rationality; all three of the latter are present in the world; and, as the commonly accepted history of Planet Earth tells us, they were introduced into the world in successive stages. No matter how much you study the fossil data you won't find the Hand of God there, but both in the fossil record and in our everyday experience we encounter the Mind of God.

Chapter

The Laws of Nature

*T*HIS IS WHERE our journey has taken us thus far:

1. The scientific method assumes that the world is understandable—and also rational in the sense that its operations can be categorized under laws and theories.

2. Science assumes that all events and phenomena have an explanation, that every effect has a cause, although in the quantum realm, cause and effect can only be *identified* at a probabilistic level.

3. The world revealed by modern science is a world that (a) obeys fundamental mathematical principles, (b) resembles computational systems with their elaborate information processing and mapping of symbols, and (c) confirms our assumption that it's intelligible and rational.

4. The laws of nature describe certain regularities in the universe. But these laws are not just descriptions of the regularities. Rather, the laws cause the regularities.

5. The laws of nature, particularly in relativity and quantum theory, can be understood and structured in the most complex and logical

thought-form known to the human mind, that of mathematics. Scientists have been stunned by the one-to-one correspondence between the "programs" of nature and the programs independently discovered and developed by the mind. Since symbolic thought and data processing are peculiar to minds as distinct from particles or force fields, it seems reasonable to assume that the laws of nature are manifestations of a sophisticated mind. No wonder then that the quantum physicist Paul Dirac said, "God is a mathematician of a very high order."

6. The paradigm of infinite Intelligence expressing itself through a hierarchy of manifestations immediately makes sense of the most diverse phenomena in our experience: rationality, intention, intelligence, beauty, and love. The denial of this paradigm comes at a heavy cost: we have to explain away the most obvious realities; the laws of nature cannot be explained and the apparent correlation between cause and effect, phenomenon and explanation is simply an illusion or at best a coincidence; there is no such thing as consciousness or intention or thought; finally, everything that seems ordered and intelligible is actually random and irrational.

So here's the upshot. There has to be intelligence in the laws of the universe or it would not exhibit the kind of rationality shown by the success of science. We're not talking of an analogy but of something actual. Our human intelligence is able to make sense of everything we observe in nature because nature itself is intelligent. Both nature and science presuppose intelligence, in ourselves, in the world and in an underlying source. Not simply the existence of the world but its thoroughgoing rationality demand an explanation, and infinite Intelligence as the ground and matrix of the world is the only satisfactory explanation. In a word, there's some underlying structure about the way the world is made. And if it's astonishing that the world exists at all, it's just as astonishing that it's a world with a structure.

We've already seen that this train of thought is neither novel nor marginal. The rationality displayed by the world, most especially in the laws of nature, has led many scientists to see it as a manifestation of the Mind of God. Einstein said it best with his stunning declaration that anyone seriously engaged in the pursuit of science becomes convinced that the laws of nature manifest the existence of a spirit vastly superior to that of men. In

Einstein's view, the belief in the rationality and intelligibility of the world that lies behind all higher-order science was implicitly belief in a "superior mind." Above all, he sought the "thoughts" of God; the rest were details. Stephen Hawking affirmed that an answer to the question of "why" the universe existed would reveal to us "the mind of God."

In "Sages and Scientists," I've already indicated that the idea of the Mind of God is not merely a metaphor. It's also the only plausible explanation for the laws of nature.

> *Geek:* I have an open mind here. But isn't it possible that the scientists who use the phrase "Mind of God" are being metaphorical?

> *Guru:* I agree that some scientists have used the "Mind of God" metaphorically, although this isn't true in the major instances. And what an odd metaphor. Why couldn't you speak of "mind of the universe" instead? It's almost as if even the metaphorically minded scientist recognizes the need to invoke a power distinct from the universe to makes sense of its workings. I, for one, am not being metaphorical. You might accuse me of being too literal. Well, I think I'm being clear instead of fuzzy. And I've consistently maintained a distinction between the intelligence we find in the world, whether embedded or active, and the infinite Intelligence that lies beyond.

Foundations of the Scientific Method

Over and over again, I've come back to the assumptions we embrace, both consciously and subconsciously, when we take up science. These are important for two reasons. They show what we really believe about the world. And they present a continuing opportunity to test reality.

I have said that these principles of science are identical with those affirming the divine Mind because in both instances we appeal to the undeniable reality of ultimate explanation, intelligibility, rationality, coherence and unification. It might be asked if any discovery has cast doubt on any of these principles. I reply that, on the contrary, every discovery consolidates and amplifies our awareness of the intelligence at work in the world. We note also that the great theories of science are popular because of their explanatory power, because they make sense of things. They confirm our conviction that there is an explanation for everything and that we can find it.

> *Geek:* I know you've already given reasons for believing that cause and effect applies at the quantum level. It still remains a fact that uncertainty and bizarre behavior are the hallmarks of quantum, chaotic and complex systems. This doesn't jive with a vision of impeccable rationality and harmony.

Guru: Well, of course there's uncertainty and indeterminacy at both micro and macro levels. And yet, as Paul Davies points out, although individual quantum events are unpredictable, collections of them conform to the statistical predictions of quantum physics. And chaotic and complex systems follow their own higher-level laws. At the end of the day, all phenomena display law-like behavior, although the kinds of laws and the levels at which they apply differ. Even when the relationship of cause and effect can't be measured or quantified, science is *always* driven by the conviction that there's an explanation for every phenomenon. This is the theme of ultimate intelligibility and rationality that is the hallmark of both science and the universe.

Here I should add that those who say that the world and life are simply random accidents are very particular about giving a non-random sequence of reasons to prove their point. Those who deny rationality in the world deploy rational arguments to make their case. Those who proclaim that purpose is an illusion are very purposeful in advocating their opinions. Those who say we're particles made up mostly of empty space still feel compelled to argue for free speech and human rights. Those who say that scientific theories are merely transient paradigms or tentative research projects don't mind flying in planes that assume the truth of aerodynamic laws.

ഌ

The Laws of Nature

Geek: What's your response to the view that the laws of nature are simply and solely descriptions of regularities in the behavior of matter? Moreover, if we find a scientifically verifiable Theory of Everything, then there seems little reason to appeal to the Mind of God.

Guru: Almost all practitioners of science agree that the physical processes and events of the world exhibit certain repetitive and predictable patterns of behavior called regularities. The regularities have been systematically catalogued into mathematical formulas called the laws of nature. Where gods and spirits were once invoked, today the laws of nature are used to explain and predict the behavior of the material world and the invariant properties of the universe. Within these laws there is a hierarchy of low-level and high-level laws. For instance a high-level law explains lower-level laws: Newton's laws explain Kepler's, quantum electrodynamics subsumes the field theory of electromagnetism. The highest-level law, we've noted, would be what's called a Theory of Everything (TOE) that describes the fundamental entities, building blocks, constants, processes, interactions and phenomena of the world in a relatively simple and unified formulation. A TOE will explain the reason why a particle has a certain mass or a force has a given strength.

As yet we have no TOE, and because of Gödel's Theorem and quantum constraints we will never have a theory that can ultimately explain *everything* in the physical world at a truly comprehensive level; this is admitted even by an optimist like Stephen Hawking. Nevertheless we do have huge libraries of laws that accurately describe the operations, capabilities, powers and properties of microverse and macroverse, particles and fields. Moreover, fundamental constants of nature, e.g. the electric charge of the electron and the strength of the weak force, dictate properties of the universe like its size.

To be sure, some skeptics have denied the existence of laws of nature. They have said that the order we see in the world is simply an illusion that the mind imposes on the world. Others have said that the laws of nature are simply emergent properties of matter. It's also argued that there are only random phenomena and interactions; any connections or correlations are purely coincidental and in any case, it's impossible to prove that a given law has always held and will always continue to hold. Such determined skepticism can only be proclaimed, not lived. Skeptics are not about to ram a car into a wall to prove that Newton's laws of motion don't work. Nor will they drink a cup of sulfuric acid in order to show that there are no so-called properties of the elements. In practice, scientists assume there are laws and that there's an underlying explanation for every event. It's not possible to do science if either assumption isn't true.

Scientific laws are not simply generated from subjective opinion. They're supported by repeated observation, successful prediction and a rationale grounded in physical and mathematical theory. Of course, all scientific theory is in principle revisable, and to this extent there's no finality to any scientific law. Scientists may try to explain certain new findings by modifications to established theories. For instance, some have claimed that the speed of light was faster billions of years ago than it is today, others that there are deviations from the second law of thermodynamics at small scales and within short timeframes. It's the weight of evidence that determines the fate of such proposals, not the question of whether or not it's permissible to modify a venerable theory.

To say there's no finality in scientific theory, however, is not to say that the principles presupposed by science, e.g., the reality and rationality of the world, are tentative. We can legitimately accept these as binding while admitting that any given theory that assumes their truth is subject to revision. The change from Newtonian to Einsteinian physics did not require a change in our conviction that events in the world can be explained. Granted, the convictions on which science is based, such as the order and rationality of the world, can't be scientifically demonstrated because they have to be presupposed by science from the start. The only sound foundation for these convictions is the meta-scientific Matrix that we discussed at the start. Rationality is fundamental to sanity and we can deny its existence only if we refuse to theorize, experiment or communicate, since all three of these activities presuppose what we deny, i.e., that rationality is real. In brief, *the theories that interpret data are subject to revision but not so the principle that underlies all theories: that the world yields data and that we are capable of collecting, organizing and understanding this data.*

It might be said that the laws and properties of the universe are accidental results of the way in which it cooled after the Big Bang. But, as Martin

Rees suggests, even these accidents can be seen as secondary manifestations of deeper laws governing the ensemble of universes. So we're still left with the question of how these deeper laws originated. No matter how far you push back the properties of the universe as somehow emergent, their very emergence has to follow certain prior laws. If, like John Archibald Wheeler, you say that the closed-loop participatory universe with its laws bootstraps itself into existence, then you're still left with the questions raised by Paul Davies: "Why *that* loop?" and "Why does *any* loop exist at all?"

Skeptics have also said that science doesn't and can't explain things; it simply describes them. Ludwig Wittgenstein once famously said that the modern view of the world is based on the illusion that the so-called laws of nature explain natural phenomena. In response the physicist Steven Weinberg points out that such skeptics mistakenly assume that "explanation" means "purpose" so that to explain something would mean discovering its purpose. This, he says, is not something science can do; science cannot tell us the purpose of the universe. But science does explain phenomena, he says, in the sense that a given physical principle can be deduced from a more fundamental principle such as, for instance, the Standard Model. Scientists still do not know which principles are the most fundamental; is it, for instance, a space-time geometry principle like general relativity or a theory like string theory? We hope that we will one day discover a unique set of laws that show why the constants of nature have to be what they are. But, says Weinberg, science can never explain the most fundamental scientific principles because an ultimate set of simple universal laws of nature would be ultimate because it can't itself be explained in terms of anything else.

It might be argued that this ultimate set of laws is demanded by some final principle of mathematical consistency. But, as frequently noted, Gödel's Theorem precludes any comprehensive self-consistency for any mathematical or logical system. And we could, in any case, ask how it is that this particular system of consistency happens to apply.

At the end of the day, then, we seem to be left with two things: the principle of rationality presupposed by science and a set of fundamental physical principles that explain all other physical principles. But this isn't the whole story because there's a third fundamental factor to be considered. This is the very extraordinariness of the fact that things follow laws at all, a mystifying phenomenon that I've highlighted from the beginning.

How does the electron know what to do? Why do all photons follow the same patterns of behavior? Why does the law of conservation of mass-energy apply across the universe? How is it possible that a material thing can follow any law whatsoever? We humans follow laws we set up because we are conscious agents capable of intentional action. But how can an inanimate thing be made to do anything, let alone follow a law?

These are questions I've already addressed, and I can't see how they can be answered if there is no Source and Matrix of the intelligence that drives everything from particles to people.

The Mathematical Connection

That anything in the universe follows a pattern of behavior is stupendous enough. But what makes this immeasurably more intriguing is the ingenuity embodied in the structure of the laws of nature.

The conformity between the highest laws of mathematics and the simplest operations of the universe has struck the greatest minds in science. Nobelist Eugene Wigner articulated this correlation most eloquently in a much-quoted paper titled "The Unreasonable Effectiveness of Mathematics in the Natural Sciences."

In recent years, Paul Davies has played a leading role in exploring the connection between mathematics and the world. In *The Mind of God* he makes the following observations:

   ◆ Physical processes can be simulated by computational operations, but even more significantly they are themselves computational processes.

   ◆ Scientific laws are basically algorithmic compressions of observable data; that is, they are algorithms for processing information.

   ◆ Mathematicians independently developed many of the mathematical models mirrored by natural processes long before the models were applied to the natural world.

   ◆ Observation and experiment do not in themselves show us the laws of nature. Rather they are clues that give us enough feedback to break the cosmic code. The laws of nature are codes that we can only discover through intelligent analysis and mathematical representation. It is remarkable indeed that the human mind has just the right ability to crack the code, when conceivably the code could have been too complicated for us. Davies observes that this ability to unveil the laws of nature brings no benefit in terms of natural selection; it is therefore not simply a biological phenomenon or an evolutionary epiphenomenon.

The fact that mathematics works so well when applied to the physical world demands an explanation. Why do the laws of nature permit physical models for the laws of mathematics?

Davies concludes that the natural world isn't just a concoction of entities and forces but an ingenious and unified mathematical scheme.

Martin Rees, whose *Just Six Numbers* was cited earlier, emphasizes the importance of mathematical laws at the heart of the universe. The sizes and masses of atoms, the different kinds of atoms and the relationships between them, he notes, determine the chemistry of our world. Moreover, the forces and particles within them control the existence of atoms.

The famous mathematical physicist Roger Penrose, co-author with Stephen Hawking of the Singularity Theorems of General Relativity, is one of the profoundest thinkers in this area. In his view mathematics uncovers truths that are already out there, truths that exist independent of us just as does Mount Everest. Penrose even calls these truths the works of God. The mathematical genius Srinivasa Ramanujam, who intuitively grasped advanced new theorems that would later be proved by complex reasoning, liked to say, "An equation for me has no meaning unless it expresses a thought of God." Another great mathematical mind, Georg Cantor, saw God as infinite, and is well known for his work on mathematical infinities. Finally, Kurt Gödel of Gödel's Theorem developed his own ontological argument for God's existence.

Talk of mathematics inevitably points to computation. We have spoken of the computational characteristics of physical processes. Paul Davies has a good deal to say on this and notes that scientists today consider the universe to be a gigantic information-processing system with the laws of nature identified as computer programs and events as the output.

The Laws of Nature and the Mind of God

The laws of nature are the laws that gave birth to the universe and keep it running. They know what they're doing, and they're the reason both science and the universe work. So where did they come from? What set up the regularities and how are different sets of regularities related?

The atheists may say *there is no reason why there are reasonable laws, no explanation for the fact that there are explanations, and no logic that underlies logical processes.* But plainly they're simply contradicting themselves. Here's what three eminent philosophers of science have said about the divine origin of the laws of nature:

♦ Atheists, writes Paul Davies, claim that the laws of nature exist reasonlessly and the universe is in the final analysis absurd. This is a view that, as a scientist, he finds hard to accept. Instead, he holds that the logical orderly nature of the universe has to be rooted in an unchanging rational ground.

♦ Richard Swinburne notes that the most fundamental theories in science are the simplest. Pursuing this line of thought, he argues that the simplest explanation for the order and rationality of the world and its laws is a free being of infinite power and knowledge. Why do material things exist with their particular powers and liabilities? Theism, in his view, provides a rational explanation centered on a being with the power and the motivation to bring all things into being. Materialism, on the other hand, does not have a similarly simple explanation for the world and also cannot explain a range of phenomena, the existence of consciousness and mind for instance.

♦ Keith Ward points out that the most satisfactory ultimate explanation of the universe, its laws and basic physical states, is a being that necessarily exists and selects basic laws of the universe in order to bring about certain good states. This explanation is supremely elegant because it unites causal and purposive explanations in one simple integrating hypothesis.

Geek: Is this the argument that the existence of law implies a lawgiver?

Guru: I appreciate your drawing my attention to a familiar fallacy. I think all theists need to purge their minds of such anthropomorphic ideas because it obscures the radical distinction between finite and infinite being. I've stressed the notion of infinite Intelligence precisely because we're dealing with an infinitely powerful Intellect free of all limitations. The laws of nature are not to be compared to rules instituted by law-giving bodies like legislatures, courtrooms or dictatorships. They're more in the nature of instruction-sets mysteriously imprinted in the very being of all things. We wonder how such a universal embodiment of intellectual power could be possible. This is mysterious beyond anything we can imagine in analogies and models.

But what we can't comprehend in terms of imagination can yet be conceived with intellect. With our minds we realize that the obvious intelligence in the world requires an explanation. The only ultimately satisfying explanation for the reality of any kind of intelligence is one in terms of infinite Intelligence, the Mind of God.

Here is how I see the relation between natural law and the Mind of God:

1. We have reviewed the characteristics of mind, its immaterial nature, its physically inexplicable generation of concepts and intentions. We have seen that matter can't explain mathematical systems or purposive structures since they both require a substrate of meaning and intention.

2. How does mind manifest itself? In the case of human beings, our minds enable us to use words. When we see a book we know it's the expression of a mind, and not just ink on paper. Likewise, all the processes in the universe bear the earmarks of mind: the mathematical codes employed everywhere in nature; the immeasurably intricate, precisely structured, comprehensively coherent patterns of interaction between energy fields and galaxies, proteins and nucleic acids; the existence of autonomous agents and purposive behavior; and the reality of conceptual thought. All of these phenomena are manifestations of mind. Scientists now take intelligence as a given, although some try to explain it away with models of self-organizing matter; this, of course, simply postpones the question one step further because we still need an explanation for the system as a whole.

3. We have seen that the correlation between the Mind of God and the rationality of nature has seemed obvious to great scientists.

The Random Option

Geek: You've made some pretty powerful arguments. But can you exclude two alternate scenarios: everything is random or there's a theory of everything that explains it all? Of these two, I find the randomness option hard to overcome.

Guru: Step by step, we're getting there, I think. Here's why I think randomness as a solution just doesn't work.

Heinz Pagels has said mathematicians have no definition or understanding of randomness; to define a sequence as random is to declare it non-random because you implicitly apply a rule. Stanley Jaki points out that complete chaos and universal chance are self-contradictory. Absolute randomness doesn't exist, he says, because every probability theory presupposes some basic regularity. As also noted, the notion that events have a random or chance origin requires us to hold that a whole infrastructure of laws has to previously be in existence. These laws are required for chance to operate. The starting-point thus is necessarily some kind of intelligence. Everything makes sense once you take intelligence as the underlying matrix.

We must remember that randomness, in any case, has no impact on the idea of God's action on the world because, as Herbert McCabe put it, God is the source of everything that happens whether it be random, free or physically caused. While science tells us why this happens and not that, God is the reason why anything can happen at all.

About the Theory of Everything as an explanation, we note first that there is no generally accepted candidate for this role. Secondly, I've pointed out that Gödel's Theorem makes any such comprehensive system impossible.

Geek: I think you've driven your point home. We've reached the Mind of God. And I think this is where you plan to move from science to philosophy.

Guru: I notice you said "We." As far as I'm concerned, we're on this trip together. I can't imagine reaching the destination alone and so I'm overwhelmed by your comment.

The last part is going to be fairly short. I've already said most of what I can usefully say about the world as a pointer to God. All that remains is to consider some questions about the nature of God. The enduring insights in this area are to be found in the works of Avicenna, et al. At best I can add a few footnotes to the *magnum opus* I've called the Matrix.

PART III
The Mind of God

*T*HE EXISTENCE OF GOD, of a transcendent eternally existent Reality that brought the world into being, has been regarded as
a self-evident truth throughout history and in virtually every
culture and society. Atheism, the denial of God's existence, is historically
and sociologically an aberration and has, from the very beginning, been as
intellectually preposterous as it was psychologically eccentric.

Geek: Before we go any further with God and atheism, I think
we should address again the question of whether we can know
anything to be true. Many modern philosophers have claimed
that we can't *know* anything. So your belief in the mind's capacity
to know the truth about things such as the existence of God is a
premise that needs to be demonstrated, not assumed.

Guru: Yes, many of today's thinkers have given up on the mind's capabilities with respect to knowing anything. So I think we need to address this
head-on. After laying out the case for the simple fact that we can and do
know things, I will also identify what I think to be two deadly sins of
modern thought, Relativism and Positivism. We call them "deadly sins"
because they're deadly for science.

The destructive impact of these tendencies was chronicled in no less a publication than *Nature* in an article titled "Where Science Has Gone Wrong." The articles indicted the four most influential philosophers of science of the last century, Karl Popper, Imre Lakatos, Thomas Kuhn, and Paul Feyerabend as "enemies of science" for whom "the term 'truth' has become taboo" and whose skepticism and nihilism "may be impairing scientific progress at this moment."

Chapter

Can Our Minds Tell the Truth?

*T*HE TRUTH THAT WE CAN GRASP TRUTHS, the knowledge that we can know, the fact that we can discern facts, these are all vital to science, theism and the Matrix. Aquinas, with his first principles, and Madhvacharya with his presentation of Sakshi laid the groundwork for the confident use of our minds.

So how do we know and know that we know? The Matrix contends that it's a fact of universal and immediate experience that human beings are not only capable of knowing but that they are indeed endowed with a knowledge-base encompassing essential and ultimate principles of reality. It's this knowledge-base that underlies all thought and rationality and every exercise of the human mind. For our discussion, we will call the distinctively human capacity to know fundamental truths about reality our "sapiential sense," from *Homo sapiens.* Our sapiential sense is a dynamic interface between thoughts and things that can discern the Actual in the Perceived and that's presupposed by all acts of knowledge. The findings of sapiential sense comprise the set of fundamental pre-philosophical and pre-scientific truths that can neither be demonstrated nor denied.

What I call here sapiential sense has been loosely called many other things: common sense, right reason, first principles, perennial philosophy, self-evident facts, the given. Since these terms have acquired too wide a range of meanings to be useful, a new vocabulary is required. The proposed phrase draws attention to the fact that we're sapient beings and will help identify and dissect this fundamental framework of human experience that's as obvious as it is obscure.

Sapiential sense is the mind's ability to see the truths that constitute reality and grasp things as they are in themselves. The seeing of these truths transcends the scope of the scientific method, which is limited to the data of the senses, and of logic, which is limited to unpacking the conclusions already contained in premises. "Knowledge," writes Illtyd Trethowan, "is basically a matter of seeing things (and) arguments, reasoning processes, are of secondary importance and this not only because without direct awareness or apprehension no processes of thought could get under way at all, but also because the point of these processes is to promote further apprehensions."

But how do we know that we do indeed know? Pause for a moment and consider some of the things of which you're certain. You know that at this very moment you're reading this sentence, that this sentence had at least one author, that the print marks on these pages contain a message, and that this message may well describe our own experience accurately.

On a higher plane, we know without a doubt that rational inquiry requires a rational order of things; that two contradictory assertions can't both be true, for example, the same object can't exist and not exist at one and the same time; that a valid inference from a valid premise leads to a valid conclusion; that the mind is capable of knowing truth, since we can't deny this statement without assuming that the mind is capable of knowing the truth of the denial; that every phenomenon has an explanation; that the world exists, that we exist (the wise response to the skeptic who denies his or her existence has been, "And who's asking?"); and, finally, that intention presupposes intelligence (here the reference is to obvious instances of intentional activity).

How do we know that these affirmations are true and how can we demonstrate their truth? Well, we know them to be true because that's what our minds tell us, instinctively, immediately, but we can't demonstrate them to be true because all demonstrations would presuppose their truth. Physical or experimental evidence is of no relevance because no amount of such evidence can show a statement such as "every phenomenon has an explanation" to be definitively true or false; a phenomenon that appears not to have an explanation may have an explanation that can't be detected with the available apparatus. A logical argument wouldn't be applicable either because the conclusions of such an argument are already presupposed before the argument gets under way.

Nevertheless we know that these affirmations are true affirmations, and if any skeptic wishes to deny them the burden of proof lies with the skeptic, though it's hard to see how a proof for the denial is even conceivable. In making such affirmations, we're remaining true to self-evident facts while those who deny them are flying in the face of these facts.

The action of sapiential sense is confirmed by our everyday experience and the collective experience of humanity. It's part of the universal heritage of the human race. Scientific activity, discoveries, inventions, theories, laws, experiments, the spirit of science, and the scientific method would be impossible without total reliance on the insights of sapiential sense. Science assumes for instance that the world exists, a meta-scientific truth discovered by sapiential sense.

Sapiential sense can't do what science can do. It can't explore the quantum or intergalactic realms. It's not a method of observation or a means of measurement. It isn't even a source of scientific theories and can't evaluate the soundness of such theories. It deals strictly with those truths that we know to be true independent of any observation or measurement and that can't be proved to be true by any amount of observation or measurement. No claim or theory that requires experimental evidence to establish it is a datum of sapiential sense.

The data made available to us by sapiential sense are:

1. affirmations
2. affirmations of such a fundamental nature that they're presupposed by, and therefore lie beyond, our abilities and activities of empirical and logical investigation; they're both meta-scientific and meta-philosophical
3. obvious and immediately known to all human beings
4. entirely congruent with ordinary experience without contradicting it at any point

5. marked by a fundamental coherence, clarity and simplicity

6. presupposed by all of our intellectual activity

7. impossible to deny without implausible rationalizations and absurd consequences

8. can only be seen to be true and are, therefore, self-guaranteeing and self-authenticating, since the truth of what is seen can't be demonstrated with external criteria.

Most modern thinkers are simply confused by the sheer quantity of data pouring in from every direction. Since it is not humanly possible to keep up with all this data, let alone make sense of it, they have retreated into relativism. What should be realized is that no amount of experimental information can negate the insights of sapiential sense. It is sapiential sense that helps us make sense of the ever-increasing influx of information. Sapiential insights provide the modern thinker with a frame of reference within which new information can be assimilated and categorized without the loss of nerve that is now a familiar feature of the intellectual scene.

Relativism

The inevitable consequence of both materialist and spiritualist monism is skepticism. Skepticism expresses itself in the popular philosophy of relativism, the idea that there's no such thing as truth since all we have are relative, subjective opinions. The *Nature* article points out that this philosophy is fatal for science: "From the false premise that all observation is theory-dependent, all routes lead inexorably to 'anything goes'...Moreover if one believes that there exists no objective truth, or even if one has doubts as to its existence, one then has no motivation, nor even inclination, to try to discover it. It is therefore highly unlikely that such a person would make any new discoveries."

Relativism is also at odds with what seems obvious in experience. Almost all human beings of the present day and throughout history have implicitly understood truth to mean a correspondence between what they assert or believe and what really is the case. Whether or not the statement "oaks come from acorns" is true depends on whether or not (a) it's a fact that oaks come from acorns and (b) this statement corresponds to that fact. To accuse someone of lying is to affirm that what they say doesn't correspond with the facts. To say that something is false is to say that it isn't a fact. This understanding of truth has been labeled the correspondence theory of truth, although it's less a theory than a description of ordinary experience.

Monism has spawned a number of philosophies built around the denial of truth. There was the coherence theory of truth according to which a statement was true if it cohered with other truths (but if we can't know the truth of any statement independently, how does the process get started?). The pragmatist theory, in its most basic formulation, held that truth is what "works" or "is useful" (but how is "works" defined and who determines if it "works?") Moreover, in our minds we instinctively draw a distinction between being true and being useful. Finally, there's the redundancy theory with its extraordinary thesis that the phrase "is true" is simply a redundancy intended to lend emphasis to an assertion.

Despite the denial of truth by many professional philosophers, it's still the case that the common experience of humanity is unintelligible without reference to the **fact** of self-evident truths. A self-evident or necessary truth is a truth that's recognized as such simply from understanding the meaning of the words we use in expressing it. In recognizing a truth to be self-evident we also realize that its denial leads to self-contradiction. "To be red is to be extended in space." "Intention presupposes intelligence." "Two contradictory statements cannot both be true," e.g. something cannot exist and not exist at the same time. We can't, like some philosophers, hunt for criteria by which to prove or verify a self-evident truth because by its very nature it's a truth that has to be evident from itself. It's self-guaranteeing and self-authenticating. Self-evident truths can't be proven scientifically, but the pursuit of science presupposes certain self-evident truths, e.g. "rational inquiry requires a rational order of things." They can't be proven logically because logic itself presupposes them. But we can use logical arguments to show that their opposites lead to absurdity.

❧

Positivism

If relativism is antithetical to science, Positivism is a view that claims to be its friend. Its starting point is what is called the verification principle: only realities or concepts that are perceptible by the senses and testable by science are real and meaningful. Since concepts like God, the soul, and morality cannot thus be verified, they are meaningless, nonsense. The main vehicle of this movement, technically called Logical Positivism, was the *Tractatus Logico-Philosophicus,* a book by Ludwig Wittgenstein. Paradoxically, another book by Wittgenstein sealed the fate of Positivism.

Although he was initially sympathetic to the ideas that the only valid approach to knowledge was the scientific one and that only scientific statements were meaningful, Wittgenstein gradually came to see that the scientific method could not be viewed as the test of truth in every field of human knowledge. He articulated this insight in his famous theory of language-games.

In this theory, Wittgenstein draws an analogy between different games and the different fields of knowledge. We play different games: tennis, marbles, billiards, football, and basketball. The concepts in these games often differ radically. A foul in basketball makes no sense in tennis; a good move in chess is meaningless in a game of marbles. There's no rule that can be applied in every game; the rules of one game aren't rules in another and often don't make sense in the other. Even the word "game" can't be said to have one meaning, for it means entirely different things in different games.

Wittgenstein applied the analogy to different fields of human knowledge. Science, theology, esthetics, ethics and economics, for instance, are varieties of language, different systems or "families" of concepts. Just as a rule in any one game cannot be taken as a rule that applies in every other game, the principles and criteria of one field of knowledge cannot be thought of as universally applicable, as rules that sit in judgment on all the fields of knowledge. It's as silly to say that the criteria of science should apply in every field as it is to say that every game should be played with, for instance, the rules of tennis. The Positivist who uses science in such a manner is playing chess with the rules of water polo. Every field of knowledge, and every aspect of human experience, it becomes clear, must be taken on its own terms, for each has its own rules. The whole idea of a conflict between science and religion is meaningless because they deal with different aspects of experience and reality; neither can nor should infringe on the other. The concept of omnipotence, for example, is not a concept in quantum physics and can only be understood and appreciated within the framework of the philosophy of divine infinity.

Wittgenstein's insightful theory helped bring about the demise of Positivism. Among philosophers, at least, "Logical Positivism" is now a derisive phrase. Sir Alfred Ayer, author of *Language, Truth and Logic,* the most influential presentation of Positivism, remarked in a B.B.C. interview that he now thought that virtually all the book's claims were false. But many writers of popular science still naively promote Positivism, unaware that its founders themselves wrote its obituary.

Positivism was a creation of philosophers. True scientists like Einstein, Heisenberg and Niels Bohr were always unhappy with the narrowness of the philosophy. Because it stifles creativity and fosters relativism and materialist monism, Positivism too is dangerous for science.

> *Geek:* I guess the die-hard relativist can continue to dispute your conclusions. But I think you've shown that, at a certain level, relativism is self-contradictory. What's more, relativism is incompatible with the practice of science. But there's a far more formidable version of skepticism and relativism around, namely spiritualist monism. This view doesn't deny simply our ability to know the truth about the world but the very existence of the world. And it's by no means just a philosophy whose time has come and gone. It's back in circulation, in a scientific context no less, in such popular books as Fritjof Capra's *The Tao of Physics.*

Chapter

Monisms of the East—Past and Present

*G*URU: I was planning to address this issue in discussing the attributes of God. But now's as good a time as any for considering the claims of spiritualist monism.

Hinduism, Buddhism, Taoism

The earliest scriptures of Hinduism, produced by the Aryans, who came to India between 1500 and 1200 B.C., are the *Vedas* (books of wisdom): *RigVeda* (path of knowledge), *YajurVeda* (path of action), *SamaVeda* (path of devotion) and *AtharvaVeda* (a book about a person named Atharvan). There is no consensus as to when the *Vedas* were composed; dates ranging from 1200 to 900 B.C. have been proposed. The *Vedas* are made up of *samhitas*, hymns used in sacrificial rituals, and *brahmanas*, commentaries on the hymns. In addition to the *Vedas*, there were interpretations of various rituals called the *Aranyakas* and theological meditations known as the *Upanishads*. The *Upanishads* are sometimes referred to as Vedanta, or end of the *Vedas*. The Hindu scriptures contain accounts of diverse deities, but commentators have noted that a distinction is drawn between these deities, on the one hand, and a supreme being that reigns over all, on the other. Prior to the coming of the Aryans, India's

ancient Mohenjo-Daro civilization had its own religious beliefs and practices. Some of these were incorporated into the Vedic religion.

The *Upanishads* are the products of different authors at different times and
do not form a single system of thought. Commentators like S. Radhakrishnan have said that they can be interpreted to fit different positions. Specific collections of texts have, in fact, served as the foundations of differing
philosophies. According to B.N.K. Sharma, theism is the normal view of
the *Upanishads.* In the earlier *Upanishads,* the authors sometimes sound
monistic when they're actually trying to talk about the transcendence and
immanence of God. One of the *Upanishads* talks of souls merging with
God like rivers becoming part of the sea. But Madhvacharya, among others, has pointed out that this analogy doesn't indicate identity; if one unit
of water is mixed with another there's union but not identity. If there was
ambiguity in the earlier *Upanishads,* observes Sharma, this was not the case
with the last of them, the *Svetasvatara Upanishad,* which is clearly theistic.
This *Upanishad* points out that human souls can't create the world and that
God, being all-knowing and omnipotent, is totally different from the soul.
It also teaches that there are three different dimensions of reality—matter, soul and God—a pronouncement that's fatal for monism. Because it's
consistently theistic in its outlook, and even gives a theistic interpretation
of monistic-sounding passages in the other scriptures, Sharma argues that
this last and most definitive of the *Upanishads* had come to the conclusion
that theism alone is the true philosophy of the *Upanishads.*

Religious works like the epics and the *Puranas* that appeared after the
Vedas were also theistic. The most famous of these, the *Bhagavad-Gita,*
affirms that the world is real and distinct from God. The word "*maya*" is
used in three *Gita* texts but not in the sense of "illusion." Geoffrey Parrinder, a doyen of comparative religion, sees in the *Gita* a "dominant and
fervent monotheism."

The profusion of religious writings and the increasing emphasis on sacrifice
and metaphysics among priests and philosophers led to a radical reaction
that came from three directions: the Carvakas, a skeptical movement that
denied the existence of the soul and dismissed the *Vedas* as self-contradictory, and two new religions, Buddhism and Jainism, which rejected the
authority of the *Vedas* and offered their own views of the world. Buddhism
is particularly important because it became India's dominant religion for
over a millennium.

Siddhartha Gautama, a warrior prince born in northern India circa 563 B.C., was disillusioned by the religious environment of his time, remote as it was from everyday concerns. After being instructed by two masters of meditation and living with a band of ascetics for several years, he spent fifty days in meditation under a *bodhi* tree. It is here that he attained enlightenment and came to know what are called the Four Noble Truths. From then he was known as the Buddha, the Enlightened One.

The Four Noble Truths are these: life is suffering; suffering springs from a chain of causes; liberation from suffering or nirvana comes if we detach ourselves from selfish cravings; and the way to liberation is the Eightfold Path, rightness of knowledge, aspiration, speech, behavior, livelihood, effort, mindfulness and concentration. Buddha rejected the *Vedas* as well as the ideas of a permanent self and an intelligent first cause of the world. He was critical of philosophy, theology and religious ritual and was less concerned with the world as such and its cause as he was with human experience.

Before his death at the age of eighty, the Buddha established a community of disciples called the *Sangha* who spread his teachings (the *Dhamma*) across the land. The new religion received a huge boost when the Emperor Asoka, who reigned from 272-232 B.C., converted to Buddhism. According to the Buddhist scholar Sanghamitra Sharma, virtually the entire Indian sub-continent was Buddhist by 250 B.C. and missionaries were sent to various Asian countries, including Indonesia, Malaysia and Sri Lanka. Subsequent kings like Kanishka and Harsha Vardhana continued the royal patronage of Buddhism. A Chinese visitor to India in 640 A.D. found that there were 100 Buddhist monasteries and 10,000 Buddhist monks in the major Indian kingdom of the time.

As is well known, there are two main forms of Buddhism: Hinayana, prevalent in Sri Lanka, Myanmar and Thailand, and the most influential version, Mahayana, found in China, Tibet, Japan and Korea. Mahayana Buddhism is for the most part monistic equating everything with an Ultimate Reality characterized as *sunya* or emptiness.

Since there have been up to 30 different schools of Buddhist thought, it is difficult to hone in on any one of them as definitive. Nevertheless, since the *Vinjnanavadins*, also called *Yogacara*, and the *Madhyamikas*, the ancestors of the Zen Buddhists, were the most influential schools, their views are relevant here.

+ Along with most Buddhist schools, they held that nothing was permanent. Everything was momentary, in flow. Also everything was sunya or nothingness, because on analysis, it is found that the things in the world are neither being nor non-being.

+ Again, most of the schools held that there is no self as such, no I, just a bundle of characteristics. The Buddhist idea of rebirth simply involves the transmission of these characteristics, not of a person or self.

+ Likewise, since there is nothing permanent, there is no God.

+ Vinjnanavadins said that the only reality is an Absolute Consciousness and that the world is not real. The latter is a creation of maya, an illusion or dream. All of our perceptions of objects outside us are false because there are no such objects. The Absolute Consciousness is beyond being and non-being, subject and object.

+ Nagarjuna (2nd century A.D.) founded the Madhyamika School. Nagarjuna held that we cannot say anything positive or negative about reality. The ideas of world, action, cause, and I are all self-contradictory. Reality is *sunya*, or void, emptiness. *Sunya* is not any one of these things: (a) being, (b) non-being, (c) both being and non-being, or (d) neither being nor non-being. This is called his four-cornered negation. Maya is the illusion that makes us believe in a real world. Maya also is subject to the same four negations as sunya. Nagarjuna rejected the Vinjnanavadin idea that sunya is Absolute Consciousness.

+ Nagarjuna held also that all knowledge is false because it involves the assumption that there is an external object, which is not the case, and a subject that knows the object, which is also not the case; therefore, "I exist" is false.

Despite its great influence, Buddhism all but disappeared from India by 1200 A.D. How Buddhism died in India is still something of a mystery. Various factors appear responsible: it was gradually assimilated into Hinduism, it lost royal patronage, or Moslem invaders destroyed the major Buddhist monasteries. Nevertheless it remains one of the dominant religions in the rest of Asia.

In its heyday Buddhism had a significant impact on Hindu thought. Among other things, the Buddhists argued for idealism, i.e. all we have are ideas in our minds and not actual knowledge of the real world, and acosmism, i.e. the universe is illusory and not real. The first major Hindu response to the Buddhists was an interpretation of the *Vedas* and *Upanishads* called the *Brahmasutras* that was authored by Badarayana in the first or second century A.D. The *Brahmasutras* speak of Brahman as the creator and sustainer of the world. Historians like S.N. Dasgupta and Radhakrishnan hold that in the *sutras* the human person is portrayed as a real agent who always remains distinct from Brahman. The author also refutes the Buddhist and monist idea that the world is simply a mental creation by affirming its reality.

But brilliant Buddhist logicians from the 1st through the 7th centuries A.D. attacked these theistic claims in the *sutras* and elsewhere. So powerful were their arguments that prominent Brahmins became converts to Buddhism while many of the remaining Hindu philosophers read their scriptures through Buddhist glasses. The first major proponent of monistic Hinduism was Gaudapada (6th century A.D.). Radhakrishnan holds that the influence of Buddhism on his work is unmistakable, while Dasgupta even suggests that he was a Buddhist. He introduced the idea of *mayavada*, the theory that the world is an illusion. Sankaracharya (8th century A.D.), the most famous exponent of monistic Hinduism, was a student of one of Gaudapada's disciples. It was Sankara, with his philosophy of *Advaita* or non-dualist Vedanta, who popularized *mayavada,* which even in his time was equated with Buddhism by critics.

In Sankaracharya's view the Ultimate Reality is Nirguna Brahman, wholly impersonal and without attributes, beyond human thought and conceptualization, beyond even the distinction of knower and known because it is "pure consciousness." There is only one thing that is real, namely Brahman. Everything other than Brahman is an appearance. The ideas that we exist separate from Brahman and that there is a world are ultimately illusions produced by *maya*. Although the world is real at one level it is not real in an absolute sense. Enlightenment comes from our realizing the identity of the self, Atman, with Brahman; once we can say, "I am Brahman," we are freed from the world illusion. The inner soul of each human being is one with Nirguna Brahman and with all other souls. The basis of Sankaracharya's teaching was his interpretation of the *Upanishads,* particularly the famous "That art thou" passage: "That which is the finest essence—this whole world has that as its soul. That is Reality. That is Atman. That art thou."

Thanks to the influence of Sankaracharya's works, *Advaita* Vedanta became the dominant Indian philosophy from the eighth through the thirteenth centuries. Although there were some defenders of theism and of realism, the view that the world really exists, like Ramanuja, their attempts were ineffective in countering the spread of monism. It fell to Madhvacharya, the founder of Dvaita or dualist Vedanta, to lead the revival of a robust and uncompromising theism.

Mahayana Buddhism and *Advaita* Vedanta were not the only monisms of the East. At least some interpretations of Taoist philosophy may be thought of as being monistic. Taoist philosophy, which goes back to at least the 4th century B.C. is, of course, very distinct in its claims and views from the Taoist religion that appeared in the second century A.D. Here we consider the philosophy, not the religion.

The author of the *Tao Te Ching* says the Tao or the Way is the source of all things but it is itself non-being or vacuous. Lao Tzu, its author, is believed to have lived between the 3rd and the 7th centuries B.C.; some scholars believe that no such person existed and the work had multiple authors. According to the *Chuang Tzu* (4th century B.C.), a sage recognizes that the life of the world is like a dream. There is a unity that underlies the apparent differences in the world: "All things and I are one." The famous commentator Wang Pi (226-249 A.D.) holds that the Tao is the ultimate reality and is beyond concepts and distinctions. Some recent commentators, e.g., Chad Hansen, have argued that the monism in the Chuang Tzu is different from classical monism, the kind found in *Advaita* Vedanta or the *Madhyamikas*. The Taoist monist doesn't accept the existence of different realities such as mind and body or God and the world. Everything is reducible to one substance, so to speak. Nevertheless the distinctions between things in the world are real and not mere appearances.

The ancient Taoist authors are famous for their paradoxical statements, e.g., "Do everything by doing nothing." Some interpreters claim these aphorisms point to hidden higher truths that can be grasped only by the initiated. Modern commentators like Hansen think that this claim is mistaken. *The Taoist paradoxes actually reject the idea that any one has special access to truth and highlight the limitations on what we can know.* If Tao is a way in which we view things, a program for action, we are told that there are no independent criteria for deciding if one such way is better than another. The Taoists wanted also to draw our attention to the influence of social environment on the ways in which we use words. Thus the injunction to not act with deliberation could be thought of as a suggestion to act naturally or spontaneously and not in accordance with the way we are trained.

The question of what is meant by the Tao in the sense of ultimate reality is ambiguous (there are some 2000 Chinese commentaries on the *Tao Te Ching*). The Tao is variously described as:

- the indescribable source of all things that is also present in everything

- non-being or the void

- nature or the world itself that perpetually goes through cycles of self-transformation.

At any rate, there is no idea or assertion of a personal God.

Madhvacharya's Critique of Monism

Without doubt, spiritualist monism is a formidable form of skepticism. Nevertheless, Madhvacharya definitively refuted it in a series of brilliant arguments. These are laid out below.

The Verdict of Experience and Reason

Sankaracharya said that difference and diversity were simply appearances caused by *avidya*, or ignorance, and *moksha*, or liberation, came from seeing the identity of our souls with Brahman. In the ultimate analysis there are no differences between things. Madhvacharya responds that difference is a primary fact of experience as judged not only by the senses but also by the *sakshi*, the experiencing self. He points out that when we know a particular thing we know it as different from other things. To know something is to know it in its individuality and in its distinction from all else. Interdependence does not eliminate difference because here we are talking of different things depending on each other.

The experience of our senses shows clearly that the world is real and that there are differences between things. Obviously our senses are sometimes mistaken. But any act of perceiving, seeing or hearing, for instance, implies that there is someone perceiving (the subject) and something perceived that is different from oneself (the object). In perceiving as well as remembering our past perceptions, we 're aware that it's we who're doing the perceiving as distinct from someone else.

The *advaitins*, of course, do not deny that we seem to perceive things. But they compare this to our experience of dreams. Although the experiences of dreams seem real, they're seen to be unreal when we awaken. Likewise, when we realize our identity with Brahman, we will see that the appearance of differences will disappear. This is because there are degrees of truth. The experience of distinction and multiplicity may be true at a lower level but not at the highest.

Madhvacharya was not impressed with these arguments. He rejected the idea of degrees of truth. Either it's the case that there are differences or it isn't. Both cannot be true at the same time. Moreover there can be no justification for denying what is obvious to experience and reason.

Madhvacharya laid out several arguments against the view that experience is illusory. For instance, suffering and enjoyment are real and never thought to be mistaken and so there's no reason why we should see them as illusions.

He also pointed out that the monist argument is self-contradictory. An illusion can come from the body with respect to something external to itself. But the self in its own inner experiences is not subject to illusion; we don't think that we are someone else. If we say that the experiencing self is itself in the grip of an illusion then we are inevitably led to self-contradiction. If everything is illusory then even this judgment of illusion will be illusory. As one of Madhvacharya's followers put it, the only illusion is the idea that everything is an illusion.

Oneness with the Absolute

The *advaitin* rejects the distinction of subject and object and says liberation comes from the realization of our identity with Brahman. Madhvacharya found the idea that we can experience our oneness with Brahman to be incoherent. To be conscious is to be conscious *of* something. This is true even of self-consciousness in which the self is subject and object. We cannot know anything if there is not something to know and someone who knows it. And if you don't have either, i.e., the object or the subject, then there can be no knowledge. The idea of pure consciousness is as incoherent as the idea of pure seeing where one does not see anything.

It's also the case that there's an infinite gulf between the human person and God. Madhvacharya contrasted human limitations with the perfection of God and said it is simply blasphemous to suggest that we are identical with God. We are aware of ourselves as limited and dependent, and our experience of our imperfection is not an illusion because our inmost self, *Sakshi*, sees it to be real and true. But God is totally independent and without any limitation. How could there be any identity between two such entirely different realities? And if we are ultimately no different from Brahman, why do we suffer the illusion of being separate? The *advaitin* says this is caused by ignorance. But who is suffering from this ignorance and when, how and from where did it come? If there is no "we," then it is Brahman who is deluded.

The Attributes of God

When we ask a monist whether or not there's a God, the reply is ambiguous to say the least. We're told that, in the final analysis, everything is one, and that the idea of a personal God is simply a primitive concept that must be left behind as we progress toward monism. The *advaitin* says that *Brahman*, the one and only reality, is beyond all characteristics and is, strictly speaking, indescribable. Madhvacharya responds to this train of thought with the simple observation that something without characteristics is no different from a void. Moreover, if something has no characteristics it makes no sense to speak of it existing at all. If we talk of it at all then we are characterizing it. If we say its existence is self-evident, then it has to be the object of an experience.

Misinterpretation of the Vedas and the Upanishads

The primary basis of the monism of Sankaracharya and *Advaita* Vedanta was a particular interpretation of the *Vedas* and the *Upanishads*. Madhvacharya authored several studies of these works to show that they were clearly theistic and were interpreted as theistic in the first commentaries.

He held that the scriptures teach the distinction between God and the world and show Brahman as the omniscient, omnipotent Supreme governor of the universe. While some verses may suggest monism, on closer study they are shown to stress the stupendous majesty of Brahman and the complete dependence of all things on it. Most important, there are other texts that teach a radical difference between Brahman and *Jiva*. Also, while the *advaitin* says Brahman has no characteristics, Madhvacharya points out that the scriptures attribute certain properties to Brahman such as reality, knowledge and infinity.

Since there are a few verses that suggest identity, e.g. "That is Reality. That is Atman. That art thou," and many others that emphasize difference, and since both views can't be true, we have to make a choice between the two options. The only viable option, says Madhvacharya, is the interpretation that is consistent with the "consolidated experience of humanity and the inference based on it," and this, he says, is one that accepts difference. The scriptures that speak of our being distinct from God are the ones that are fundamental while those that suggest identity are figures of speech highlighting intimacy and dependence. In fact Madhvacharya shows that all the usual texts cited by the monists actually make the case for an absolute distinction between God and the world. And the *Svetasvatara Upanishad*, the last of the *Upanishads* that presumably is also the last word on the subject, is undeniably theistic. Faced with the theistic texts, the *advaitins* have said the many parts of the *Upanishads* that teach distinction are "non-truth-declaring." But this is a case of wanting to have their cake and eat it too. If the monist says these verses are false then that undermines his/her own position since it is based on the veracity of the scriptures.

Like others before him, Madhvacharya made the case that *Advaita* Vedanta was actually Buddhism expressed in the language of the *Vedas*. For instance, he says, a Brahman without attributes is no different from the *sunya* or Void of the Buddhists. Other similarities include the ideas of *maya* as illusion and *nirvana* as liberation from ignorance.

Given the disputable basis of Sankaracharya's *Advaita* Vedanta, and the power of Madhvacharya's critiques, we might ask why *Advaita* has become so dominant as to be identified with Hinduism. Two historical factors are responsible:

1. Although Madhvacharya established a powerful school of

thought with intellectually noteworthy followers, his works were all but unknown to most of the English-speaking world until the 20th century.

2. Meanwhile, Sankaracharya and his philosophy received a huge boost through the efforts of popularizers like Swami Vivekanda. This established brand recognition.

But none of this changes the fact that the monistic teaching of *Advaita* Vedanta is fatally flawed because:

♦ Sankaracharya's monism is based purely on his interpretation of certain texts in certain books.

♦ His interpretation goes against experience and reason and he ignores texts that deny his theory.

♦ Madhvacharya and his successors have created a formidable arsenal of arguments against monism that have yet to be answered.

The Physics and Philosophy of the Tao

As you've noted, there has been a fresh outbreak of spiritualist monism in scientific circles. A profusion of popular science books, the most famous of which are *The Tao of Physics* by Fritjof Capra and *The Dancing Wu-Li Masters* by Gary Zukav, propagate the idea that the findings of the New Physics establish the claims of Eastern monism.

Although these books were sensations in their day, scientists and philosophers have evaluated their basic claims and declared them gravely deficient. Jeremy Bernstein, for instance, observed that Capra's book is not even wrong. The problems with these books are in two areas: their portrayal of Eastern thought and their interpretation of modern physics as it relates to Eastern ideas.

Capra's agenda is clearly monistic. He deplores the divisions created by Western science between spirit and matter, heart and mind, and calls for a return to the supposed insight of the East, the realization that we are all part of the same Ultimate Reality. He notes that quantum physics eliminates the distinction between the observer and the observed, and introduces consciousness into science. Like Eastern mysticism, modern physics shows the world to be unified and essentially unseparated.

As it relates to Eastern thought, an obvious weakness in Capra's book is the assumption that Eastern religions are monolithic in their fundamental views; he says the basic features of their world-views are the same, a questionable claim. By Eastern religion, he refers primarily to Hinduism, Buddhism and Taoism. It should be clear from our discussion that not all Eastern religions are monistic as Capra assumes; in fact, only one of six strands in Hinduism is monistic. The physicists George Sudarshan and Tony Rothman highlight another such discrepancy. Capra associates the scientific interest in symmetry with the Greeks and notes that this non-dynamic idea of symmetry doesn't play a role in the East. But Indian spiritualists are concerned with spiritual symmetry and the Chinese place a high premium on equilibrium and balance.

The historian of philosophy Frederick Copleston points out that the idea of the unity of the world really springs from Eastern thought and not mysticism as Capra suggests. Secondly, it's misleading to compare the essential outlooks of Eastern thought and modern science. The focus on the unity of all things in Taoism was centered on the question of how human beings should behave. Buddhism was a program of spiritual advancement. Modern physics, in contrast, is not part of a program of spiritual life. Finally, Eastern philosophies thought of the One as being more than the sum total of the Many, and this is hardly compatible with Capra's thesis that in modern science the world in itself is the ultimate reality.

On the physics side of things, the obstacles are no less serious as shown in two major articles: "Eastern Mysticism and the Alleged Parallels with Physics" in the *American Journal of Physics* and "Parallels and Paradoxes in Modern Physics and Eastern Mysticism" by Sal Restivo.

In Capra's view, physics is driving us to monism; consciousness has come to play as pivotal a role in physics as it does in mysticism; the subject-object distinction has been eliminated; and the idea of fundamental laws of nature must die. But each one of these claims is objectionable.

The unity of all things that we find through quantum interactions is said to prove that all things, including ourselves, are simply manifestations of the same underlying reality; we are All-in-One. But the relatedness shown by science is by no means evidence that we're all just part of the same thing. In fact two things can be related only if there's some difference between them. Such unity can more plausibly be seen as evidence for an intelligent

Mind as the Source of all things. It's precisely the unity of fundamental interactions, constants and laws that leads us, like Einstein, to applaud the rationality incarnate in nature.

Often enough, the parallels are purely linguistic. Restivo points out that physicists and Eastern philosophers sometimes use the same words but with entirely different meanings.

Consciousness is one instance of such confusion. The importance of consciousness in quantum physics is seen as validating the role of consciousness in Eastern thought. But consciousness is understood in radically different senses in mysticism and quantum theory. Whereas the consciousness spoken of in the East is a fruit of *detachment from the senses*, in science we speak of it when considering the awareness of events that comes *through the senses*. The quantum physicist John Stewart Bell, whose Bell's Theorem is a mainstay of the parallelists, holds that "it is not right to tell the public that a central role for consciousness is integrated into modern atomic physics."

Schrödinger is the scientist most often cited as seeing the connection between monistic mysticism and science, particularly with respect to the idea that there's no distinction between subject and object. But Schrödinger explicitly denied this monistic view: "It is maintained that recent discoveries in physics have pushed forward the mysterious boundary between subject and object. We are given to understand that we never observe an object without its being modified or tinged by our own activity in observing it... Still I would not like to call this a direct influence of the subject on the object. For the subject, if anything, is the thing that senses and thinks. Sensations and thoughts do not belong to the 'world of energy.' They cannot produce any change in the world of energy."

Capra decries the idea of fundamental laws of nature. He acknowledges the theistic origin of this idea. These fundamental laws were seen as laws instituted by God, eternal and invariable. In the East, he says, no one thing is more fundamental than another. Unity and inter-connection are the main themes. The property of each part determines the properties of every other part. So the notion of fundamental laws, constants and equations must be replaced by the idea of a network of relationships. The very idea of foundations of science must go. Capra's hypothesis "accepts no fundamental entities whatsoever—no fundamental laws, equations or principles."

With this last recommendation, Capra highlights the extreme danger of monism for science. After all, the idea of fundamental laws played a major role in the formation of modern science. In fact, it's hard to imagine practicing science without some sense of laws of nature. Are we to assume that Newton's Laws of Motion don't apply to all instances of normal motion? And what about the Theories of Relativity and Heisenberg's Uncertainty Principle? To do away with fundamental laws is ultimately to do away with science.

Restivo concludes his critique of the Capra-Zukav program with the observation that evidence and logic do not support the idea that Eastern mystics anticipated modern physicists. Consequently, the supposed parallels between their respective enterprises seem spurious.

It might be said that what Capra is doing is simply the monistic equivalent of my attempt to relate science to a theistic matrix. But there's a key difference. Capra's approach is one of parallelism and analogy. The Matrix claim, on the other hand, is about foundations. Specifically, we're unpacking the premises of modern science and showing that they can only be justified with the rationale provided by the Matrix; furthermore, the very themes defended by the Matrix served as the motivation for science. I certainly don't wish to claim that ancient mystical texts confirm the messages of modern science.

Monism has also entered modern thought through the back door via interpretations that are not directly linked to monist philosophies. A prominent example, considered earlier in our discussion of space and time, is the best-selling book *The End of Time* by the physicist Ian Barbour. Barbour's thesis is that change and time are both illusions.

It hardly needs to be said that spiritualistic monism is inimical to the aims of science. While science seeks unity, science cannot accept the idea that everything in the world is a single thing, the ONE. Nor can it brook the idea that diversity and change and the external world as a whole are illusions. Science assumes that energy fields are real and physical and manifest different kinds of behavior. Science assumes you can quantify and measure. Science also assumes there is a self, distinct from the world that can and does learn more about the world. For all their talk of "consciousness," the New Age populists seem positively hostile to the idea of self-consciousness. But how can consciousness and intelligence exist without a center, a locus? It's simply incoherent to talk of intelligence if there isn't someone who's intelligent or consciousness if there isn't someone who's conscious.

Chapter 3

From Cosmos to Theos

I'VE TRIED TO CHRONICLE the errors of both materialist and spiritualist monisms. It's time for me to complete the case for my own position, theism.

In earlier presentations, I've outlined three ways in which we can come to recognize the existence of God. Three hard facts emerge in our immediate experience: (1) we are rational, conscious beings; (2) we live in a world of intelligent systems with the primary embodiment of this intelligence in the laws of nature; (3) things exist and none of these things can explain how they came to exist without reference to a pre-existing cause. These three facts led to three conclusions that seem to me obvious:

1. The intelligence that's either embedded (an electron) or active (an organism) can only be explained by reference to the existence and immediate activity of infinite Intelligence. Inert matter cannot produce an intelligent system given even infinite time.

2. Minds can only come from an infinite Mind. Consciousness can only arise from an eternal Consciousness. It's simply incoherent to suppose that matter, blind, mindless matter, could ever produce consciousness or thinking.

3. Absolute nothingness cannot produce something given endless time; in fact, there can be no time in absolute nothingness. Since

the something that exists in the world does not contain within itself an explanation for its own existence, it can only be explained in terms of a Being that explains both its own existence and that of everything else.

> *Geek:* I must say that your arguments for a Conscious Mind as the source of conscious minds and a higher Intelligence as the source for intelligent systems are plausible, even compelling. But I think your third argument is pretty weak. Even if I grant that nothingness cannot produce something, I can plausibly say that the something that exists always existed, obviously in different forms but as something.

Guru: As always, I appreciate your candor. For my part, I find the third approach as compelling as the first two and even more fundamental. I'll try to show you why.

We constantly seek explanations because we perceive intelligibility and rationality in the universe around us. This pursuit of explanation reaches its climax when we ask how it is that anything exists at all. "Why is there something rather than nothing?" asked Leibniz, and here the "why" question is less about ultimate purpose than it is about ultimate origin.

The question is equally forceful whether the universe has hundreds of billions of galaxies or whether all that exists is a grain of sand. How did it, the galaxies or the grain of sand, get here? It is now in being. How and why was it brought into being? Did it create itself? Or was it always here with no beginning and, presumably, no end? But to say that it always existed is still not an explanation of its existence. It is, rather, a highly speculative description that is, by the nature of the case, incapable of scientific verification; how is it possible to prove that matter and energy have no beginning and no end? And a description is not an explanation. The assertion that the universe always existed is not, to my mind, an explanation for its existing; even if we admit the assumption of an eternally existing universe we are still left with the problem of explaining and accounting for the phenomenon of an eternally-existing universe.

The mathematics and the mechanisms behind the processes that culminated in the universe we inhabit have been the objects of plausible and often fruitful speculation. But the question of ultimate origin, an ultimate explanation for the mathematics and the mechanisms, continues to elude and baffle the most ingenious theorists.

Science takes as its starting-point the meta-scientific Principle of Explanation. According to this principle, (a) reality is intelligible and rational, (b) there is an explanation for everything, and (c) any thing that exists must have an explanation adequate to fully account for its existence either in itself or in something else. Once we recognize this Principle of Explanation, on which all of science is built, as an essential and ultimate principle of reality, we are led inescapably to the existence of a Supreme Being that is the ultimate explanation of all that exists. Let me try to show you why this train of thought is inescapable.

The universe is made up of things, e.g. fields, particles, the spatio-temporal continuum, stars, plants, which do not explain their existence but can only be explained by reference to some other entity or law. The universe itself is simply a label we give for all of these things; there's no super-entity called the universe over and above the things that constitute it. Now a thing that doesn't explain its existence contributes zero to its own existence or coming to be. It owes its origin to other things. The existence of any one of the things in the universe poses a major problem. It has a cause; someone or something else brought it into existence, and this cause is also caused by still another cause, which was also caused by yet another cause, and so on.

From a rational standpoint, there are only two possible explanations for the existence of this particular thing or any of the other caused things that constitute the universe. Either there's an infinite series of caused causes or there's an ultimate and primary uncaused cause, a cause which doesn't owe its origin to any one or anything else and can fully explain its own existence. The existence of an infinite series of caused causes would still not explain the existence of even a single caused thing; every being or thing in the infinite series contributes zero to its own origin and, as we've seen, an infinite number of zeroes is still zero. Just as an infinite series of postal carriers can't explain the existence of one letter, an infinite series of things that were caused can't explain the existence they've received. Let's say that a scientist says that a particular geometric manifold or symmetry group is the starting-point. We're still left with no answer for our question. As William Stoeger notes, there is nothing in the nature of either geometric manifolds or symmetry groups that compels them to exist or explains why they exist. They need an explanation for their existence as much as anything else in the world.

In short, any thing that exists must have an explanation adequate to fully account for its existence, either in itself or in something else. The only viable explanation for the existence of any one of the entities or all of the entities that make up the universe would be the existence of an ultimate uncaused being, a being that did not receive existence from anyone or anything else and can completely explain its own existence. This self-explanatory being is commonly called God, and is the explanatory ultimate demanded by all non-self-explanatory entities from sub-atomic particles to galaxies. We do not reason from the fact that everything in the universe has a cause in space and time to the conclusion that the universe has a cause in space and time; rather, everything in the universe is non-self-explanatory, which means that the explanation of the universe does not lie in itself but must lie in a self-explanatory being.

The Cause of all things is not a cause like the physical causes we witness in the world. It's a transcendent Cause that's qualitatively different from all other causes. I'm aware that the idea of cause is used in many different senses, from the interaction that leads to the collapse of a wavefunction to matter curving space-time to the free choices of intelligent agents. But all of these ideas of cause in some sense or other revolve around the notion of providing an explanation for an event, thing or phenomenon. Even in our immediate experience we see different kinds of cause at play. We see physical bodies acting as physical causes. But when we decide to lift our hands, our minds instruct our brains to do so. A trans-physical cause brings about a physical effect. There's no violation of the law of conservation of mass-energy since any physical exercise of a mental choice will take place in accordance with this law. When it comes to God's creative activity, it's obvious that a Cause that brings something into being from nothing is radically different from the causes at work in the world. The latter kinds of causes receive their very nature and efficacy from the primordial Cause.

> *Geek:* I can think of at least five different issues that need to be resolved.
>
> 1. Who created God? There's no point saying that God is by definition "uncreated" because that's simply a play on words.

2. Second, why can't we just say that the universe just happens to be here, end of story? It's meaningless to ask how it got here once we've figured out the fundamental processes and laws that govern its structure. The so-called Principle of Explanation only applies to material processes.

3. Thirdly, the physicist Edward Tryon has shown that the net energy of the universe is almost zero and so there's no contradiction in saying that it came to be out of nothing since it is nothing. If you add up the binding or attractive energy of gravitational attraction, which is negative, and the rest mass of the whole mass of the universe, which is positive, you get almost zero. No energy then would be required to create the universe and therefore no Creator is required.

4. Fourth, how do you know that the Creator is infinite and perfect? How is it possible to know anything about such a being?

5. Finally, there's an absolute disconnect between the existence of a perfect Being and the prevalence of evil and suffering in the world. If there were an infinitely powerful Being, then it would have prevented evil. Since evil is present either there is no such Being, or the Being is not infinitely good.

Guru: All of these are reasonable questions and as you well know they've been addressed time and again, and adequately so in my opinion. But it's worth retracing our steps along this well-trodden ground if only to update ourselves.

About the question who made God, I've already said that it betrays a misunderstanding of the concept of God. This is not a play on words. The problem is actually conceptual confusion on the part of the questioner. If the universe is said to be ultimate, it's inexplicable because its existence *cannot* be explained. If God is taken as ultimate, God's existence *need not* be explained because God is self-explanatory. This distinction was well put by the particle physicist Stephen Barr in his 2002 Erasmus Lecture: "The mystery is not impenetrable to intellect or unintelligible in itself; rather, it is not fully intelligible to us. And reason itself tells us there must be such mysteries." As Hugo Meynell has said, God is that on which all other things depend for their existence and so God cannot be thought of as depending on anything else. To ask who created God is to ask who brought into existence that which by its very nature always exists and cannot not-exist. But, you might ask, why should we believe that there is such a being? This leads to your second contention.

Is "how is it that the universe exists?" an illegitimate question? This was the claim made by the skeptic Bertrand Russell when he said in a debate that "the universe is just there and that's all." We shouldn't ask for its cause since causes apply only to the things in the universe and not the universe itself. What he and you are saying here is that the universe is the ultimate brute fact. There is no explanation for the universe and we shouldn't expect any explanation for its existence. But this to me is simply a dogmatic demand to cease rational inquiry, a demand that is incompatible with both reason and science. Why shouldn't I continue my explanatory quest all the way to the start of the process? Why shouldn't I continue to ask why and how to the very end? This quest is powered by nothing less than the innate dynamic of reason and understanding. It seems to me that any attempt to halt these rational processes rises from fear of the inevitable conclusion. The train of thought that leads from the universe to a transcendent self-explanatory Being is, above all, the response of rationality to the facts of experience. Because it rises in the "movement of reason," theism is the "ultimate rationalism."

It has been pointed out that those who make comments like the universe is "just there" (Russell) or it's "gratuitous" (Jean-Paul Sartre) betray by their very choice of words the dissatisfaction of reason with such a state of affairs. Science cannot rest with brute facts, facts that are neither intelligible nor explanatory. In fact no true scientist will say this is how things are and we shouldn't ask why and how. Now God as the explanatory ultimate is not a brute fact because here we're talking of a Being of infinite Intelligence, the Source and summit of rationality. God is the Cause that makes sense of all other causes, the Reason that makes reason reasonable, the Explanation of all explanations.

Regarding Tryon's gambit, the atheist J.J.C. Smart, in his debate with John Haldane, points out that the postulation of a universe with zero net energy still doesn't answer the question of why there should be anything at all. Smart notes that the hypothesis and its modern formulations still assume a structured space-time, the quantum field and laws of nature. Consequently, they neither address the question of why anything exists nor confront the question of whether there is an atemporal cause of the space-time universe.

Your last two questions require fuller responses and I will now turn to these. First I will address God's infinite perfection and second I will consider the problem of evil.

Chapter 4

Infinite Perfection

*T*HE RECOGNITION OF THE EXISTENCE of an infinitely perfect Being that always existed represents the fulfillment of human rationality because it makes sense of all levels of our experience. But it also heightens our awareness of the mystery of existence; we ask how it is that God always existed. This, if you recall, is the question answered by the Matrix. To recognize that our existence is unintelligible if it is not grounded in a self-existent God is also to recognize, without knowing how this is possible, that God exists always and that His Existence is free of all limitations. This is the idea we will now explore.

Let's start with a what-if thought experiment. What if there's a being that always existed? How could this be possible, we ask. How could something *always* exist? What properties would such a being have? We know existence is not a property like being red or round or good. You have to exist to have properties. But here we are talking of a being with unlimited existence, one that always exists. It not only exists without end but also has no beginning. *"To be" is the very essence of God whereas it is possible to conceive "not to be" of every other thing.*

If it's your very nature to exist, if your inner being requires that you always be, then you enjoy existence without limitation, what we might call the fullness of existence. But to exist without limitation is simultaneously to possess all the perfections of existence without limitation.

"To be" is to be a certain thing that possesses certain perfections. The first perfection for any entity is that it exists, and as we go up the ladder of life we encounter other perfections such as intelligence and will. But none of these other perfections would exist if a being that existed did not possess them, and so they can rightly be thought of as perfections of existence. If, then, to exist is to possess in different degrees the perfections we see in the world, then to exist without limitation is to possess all these very same perfections without any limitation.

This basic insight was well articulated by H.D. Lewis, who said that whatever can serve as the ultimate ground of all else must be thought of as complete and adequate in itself in every way; beyond everything we encounter in the world there is a transcendent reality which, by the nature of the case, must also be supremely perfect.

God is thus:

- ♦ a Spirit, present everywhere by acting everywhere, not limited by matter or space

- ♦ eternal, seeing all things in a single everlasting present without any succession of events, not limited by time

- ♦ immutable, not susceptible to any change

- ♦ omniscient, knowing all there is to know as Creator of all things, not limited in knowledge

- ♦ omnipotent, capable of doing anything that does not contradict logic or the divine nature; to make something from nothingness is only possible for someone with infinite power, not limited in power

- ♦ perfect; to be imperfect in any respect is to have imperfect existence

- ♦ simple; the divine perfections cannot be ultimately distinct from each other or from God. As shown by thinkers like Eleanor Stump and Norman Kretzmann, what we describe as God's omnipotence or His omniscience are just different ways for finite minds to describe a single eternal action or different manifestations of the action, this single action being God himself.

And there can be only one such reality, only one God, because if there were more than one such being, then we still would not have found an ultimate and final explanation, since we would have to find an explanation for the relationship between these beings. Moreover, the different beings would not be infinitely perfect because each would lack the perfections of the other(s). Thus, by the very nature of the case, there can only be one infinitely perfect being that is a truly ultimate explanation of all other being.

Now sometimes when we speak of God as a being we seem to imply that he is one more entity in a class of beings. But when we call God a being we simply use an analogy from our experience of finite beings. We should caution ourselves that he is not a member of a class or thing among things. He transcends all classes. He is the infinite Source of all finite things, the ground of their existence.

Chapter 5

Before Time and Outside Space

*T*o say that God is changeless and eternal is not to say that God is static. On the contrary, God is infinite energy and activity and, as Illtyd Trethowan points out, this activity, being infinite, cannot be either less or more than it is.

Whereas experience is inevitably successive in nature for finite beings, and time is the measure of such change, God has no series of experiences but possesses all perfections in a permanent present. Boethius described this everlasting present, this never-ending Now with no beginning or end, past or future, in his classic definition of eternity as "the utter and complete possession of life without end in one simultaneous act."

The difficulty most people have is not with the concept of eternity as such, mysterious as this will always be, but with the relation of God's eternity to time. Since God is totally outside the world of finite being, and since time is a property of finitude, human lives, the whole of cosmic history, and all time, are eternally present to him, present to him in his never-ending Now. What is yesterday, today and tomorrow, is TODAY to him.

The fact that we perceive God's timeless actions from our time-limited perspective does not imply that he is involved in time; it only shows our limitations. While we cannot be rid of these, the use of analogy can help us in understanding God's eternity as it can all the other infinite attributes.

There's no such thing as foreknowledge for God since what appears to be foreknowledge to a human person is part of God's eternal knowledge. To give an example, God eternally sees Julius Caesar's birth, upbringing, military campaigns and death, each circumscribed by attendant geographical and historical conditions. But God's vision is not thus circumscribed. Although God's eternal NOW is inseparably linked to the successive, circumscribed "nows" of history, it's not positioned along with them. Thus if we're asked whether God now at 6:00 A.M., Tuesday, knows a free act one is going to do tomorrow at 6:00 P.M., it can be replied that within our timescale it's impossible to know an act which hasn't yet occurred and doesn't exist. But God in his eternal NOW knows always that free act because he sees it as it is *actually occurring* and not in the future. But his NOW cannot be placed alongside any created "nows."

Inevitably, those who exist in time experience God's eternal activity in terms of time. As emphasized earlier, our experience of time isn't an illusion and certainly we shouldn't think that God doesn't see our experience for what it really is. God timelessly sees the things of time as being in time. This implies neither that they're not actually in time nor that their time-bound nature affects the timeless nature of God's perception of them.

One of the most significant recent works on God's eternity is Brian Leftow's *Time and Eternity*.

God is outside space in the same way that the novelist is outside the world of his novel, says Leftow. Again, though God is not located anywhere in space, he is *next* to everywhere. God's relation to those who dwell in the space-time web may be likened to the relation of the three-dimensional interior of a sphere to the points on its two-dimensional "skin." An analysis of God's basic actions shows that in his very act of creating the universe, God wills its every state of affairs.

Another way of looking at God's acting in space is by reference to our own experience. Although the human self is not physically extended it is present where it acts; the self is not in the thumb but when you consciously wiggle the thumb, the self is acting there. Likewise, God is present where he acts, and since he holds all things in being, he is present everywhere by virtue of his conserving power.

With regard to time, Leftow argues that God must exist outside time since to live in time means that different parts of one's life come to an end, a clear imperfection. To say that God's life is timeless is also to say that he sees the whole span of time at once, lives his entire life always, does all that he ever does at once and does not change in any way in what he knows or intends. There is no monotony in his timeless existence, says Leftow, because the infinity of the truth he sees and the perfection of his grasp of it, the infinity of his knowledge and love and the infinite Good that he is in himself, necessarily results in a state of the greatest possible joy in his timeless and "unimaginably intense" present.

Chapter

The Problem of Evil

*T*HE PROBLEM OF EVIL in the world is certainly the most serious issue in any discussion of God. But it seems to me clear that the existence of evil does not in any way pertain to the existence of God: it is relevant rather to the nature of God. We might ask if he is all-good and all-powerful given the existence of evil.

I say that evil doesn't impinge on God's existence because regardless of the evil in the world, we still have to explain three things: the fact that the world exists, the intelligence in the world, and the reality of conscious thought. The greatest evils can't erase questions of origin, and these origin questions point clearly to an eternally existent, infinitely intelligent being.

But once we recognize the inevitability of God's existence, we're baffled by the fact that there's evil in the world he created. While I admit that no theoretical explanation of evil and pain can alleviate its misery, I still think that we have some idea of why there is evil.

One of the chief attributes differentiating us from inanimate matter and animal life is our ability to choose freely. If we could not make free choices we could not love, we could not create and achieve, we could not be all that makes us human. A union of love between God and the human person is possible only if the human person is free, free to choose for better and for worse, free to accept or reject God. But if freedom is essential to our being human, it is necessarily a double-edged sword. To be human means that you can also make bad choices and act maliciously. Being human means being capable of the Holocaust, the killing fields of Cambodia and the savagery of 9-11. In short, if God has indeed created us free and if freedom necessarily involves the possibility of choosing either good or evil, then the exercise of freedom is responsible for the evil in this world, and God permits the inescapable option of evil in order to bring about the greater good of enabling us to freely love him.

Sandra Menssen and Thomas Sullivan have argued that the question of why God permits evil doesn't have to be answered in order to resolve the apparent incompatibility between the existence of a good God and the existence of evil. In order to speak about a problem of evil, we must have criteria specifying what constitutes a "good" world. In their view, there are four possible criteria for judging the "goodness" of a world: functional, utilitarian, aesthetic, and theistic. It may be said that whether a world is good or bad will depend on whether it fulfills its function. But a world can have a function only if it exists *for* something, and such a function would have to be given by a Designer. Utilitarian criteria for the goodness of a world face various criticisms, such as, who is to decide what is beneficial to the utility of the world as a whole? And if the world has an infinite Creator, utility for him comes before utility for anyone else. Aesthetic criteria cannot suffice to satisfy our general criteria of good and evil. On this basis, Menssen and Sullivan conclude that the only viable criterion for judging the goodness of the world presupposes the existence of God, so God's existence is acknowledged before the problem of evil and suffering is even addressed.

I grant that none of this will comfort those distraught at the loss of a loved one or languishing in physical agony. To them I can only say that the infinite love of the God who created them for union with himself is the sole source of ultimate hope and consolation. This is a love that we experience here and now and can enjoy forever if we choose a destiny of union with the infinite Lover.

One final thought. We hear a lot about the problem of evil. And yet the mystery that is infinitely more paradoxical and puzzling than the problem of evil is the existence of truth, beauty and love. This is what I would call the problem of good, and it can only be explained by the existence of a Being Who is nothing less than the fullness of Truth, Beauty and Love.

Evil is not a creation of God. It's our exercise of our negative power to refuse the good. It is a corruption, misuse or perversion of the good for which we alone are responsible. It resides in our intentions and choices. It resides in us. And an answer to the problem of evil, if there is to be any, can only be found in the mystery of human freedom.

> *Geek:* Your arguments about the roots of human evil are credible. But how about the suffering of the innocent that has natural causes, floods, diseases, famines and the like? What possible rationale can there be for allowing any of this to take place?

Guru: The problem of pain and suffering caused by natural phenomena is in many respects more troubling than the problem of moral evil. In the case of evil, we know that it's a result of free actions and hence plausibly it's not something for which God is, so to speak, responsible. But the reality of babies with brain tumors and large-scale loss of life caused by earthquakes and tidal waves is another matter altogether. No human agent consciously brings these about. Many have said that God, being all-powerful and all-knowing, could and must prevent such tragedies if he's all-loving.

Various approaches have been adopted in solving this problem. Some say that evil, pain and suffering are simply illusions. Others say that natural calamities are the results of the free actions of supernatural agents, bad angels, for instance. A third view holds that God allows these to enable our spiritual and moral growth. None of these solutions seem to me adequate. Certainly, pain is no illusion since there's no doubt that we *feel* pain and the feeling at least is no illusion. And the claim that supernatural agents cause calamities still leaves open the question of why innocent humans have to suffer as a result. The argument that God brings about suffering from natural causes in order to mold us also faces problems: what benefit can follow from the painful death of a baby?

May I suggest the following approach?

1. God has instituted certain fundamental physical laws in the universe. If no laws existed or if God constantly interfered with these laws, we would live in a world of chaos. There could be no free human actions and no history.

2. To live in a world where such laws are in place means having to live with the cause-and-effect consequences of these laws.

3. Sometimes these laws result in suffering for humans and animals. If cells become cancerous or if movements in the underlying astenosphere apply pressure on the earth's outer shell, then the end result is calamity for those affected.

4. If there is no interaction of cause and effect in the physical world, there could be no world order. The physical dimension of our existence necessarily entails some level of subjection to physical laws.

5. God permits such pain and suffering because some greater good will result in the end. Although we do not know all dimensions of the greater good, we see some traces of it in this life itself. But ultimately, I believe, a final resolution of the suffering of the innocent is to be found in a life beyond death. Almost all the major religions have affirmed a life after death that is related to our choices and actions in this world.

Geek: But couldn't God have created a world in which the laws of nature were different, where there could be minimal suffering brought about by natural causes?

Guru: Clearly, any physical world will require physical laws. The laws we have in our universe are an essential part of what makes us the kind of beings we are. Speculation about other possible worlds is a favorite pastime of philosophers. Scientists have gone down this path as well with various multiverse theories. But my concern is with the world in which we actually live. Is this the best possible world? I don't believe there is any such thing. For every version of a perfect world we can think of another that is better or has a different twist.

But we do know that a world with free moral agents is better than any world made up of automatons. And moral evil is an inescapable option in the context of freedom. Even an omnipotent being cannot make a free agent morally good. To be truly free, the agent must be capable of making choices with no compulsion from God. Thus any world with free agents necessarily opens the door to evil.

Does this mean that we also have to be subjected to the harshness of nature? While the idea that the world is a "vale of soulmaking" has its limitations, it seems obvious that we couldn't become courageous or self-controlled or diligent if we didn't have challenges and trials. And natural laws are a basic component of the kind of world order required for the development of free moral agents like ourselves. Free will would be impossible, observes Gerald Schroeder, if God were to intervene regularly once the laws of nature were in place. But to return to the original question about the suffering of the innocent, a final answer to the problem of physical evil can only be found in a life beyond death. The great world religions tell us that this is one of the greater goods brought about by God despite the evil, pain and suffering of this world.

All I said earlier about the mind shows that there's something about us that cannot be reduced to the physical. Consequently, some kind of survival after death is at least conceivable. But this is a topic that would take us to a whole new series of dialogues. Let me simply conclude here by noting that an ultimate solution to the problem of evil, in my view, involves the possibility of an after-life.

Chapter

The Denial of the Divine

*G*EEK: As I've indicated, on the whole I find your arguments to be credible. But it's clear that many brilliant thinkers have rejected them. Many famous philosophers and scientists are atheists. On what basis can you claim to be right when sophisticated thinkers take a very different view of the very same issues?

Guru: It's a truism that philosophers disagree on almost everything. It's equally plain that atheism and skepticism, here understood as the rejection of belief in supernatural agents and interventions and in an after-life, and relativism, the belief that we cannot know anything to be true, represent the establishment position in modern thought. And to those who are intellectually intimidated by the spectacle of scholars trotting out seemingly conclusive arguments on both or multiple sides of most issues, skepticism and relativism offer the paths of least resistance.

Historically, skepticism and its offshoot atheism began in the Western world with such Greek thinkers as Epicurus (341-270 B.C.) and Lucretius (94-55 B.C.) and remained an intellectual force until about 450 A.D. The skepticism of this era was not necessarily a reaction to religious belief and co-existed with various other world-views. From 500 A.D. until the sixteenth century, monotheism was dominant in the West. It was in the late sixteenth century that skeptical views again returned to prominence.

Initially, the skeptics rediscovered the works of the Greek skeptics and rejected, either directly or indirectly, the Christian religion while remaining deists, although some of the French skeptics rejected even deism. Their skepticism was at least in part a reaction to Christianity. By the nineteenth century, skeptics no longer held on to deism and sought to explain religious belief in secular, e.g. psychological, sociological, and economic, terms. In the twentieth century, skeptics turned to full-blown atheism without reference to the demands of religious belief-systems. Today, in fact, skepticism is the mainstream, establishment position of modern academia in most parts of the world.

The intellectual foundation of atheism and skepticism, understood as a rejection of belief in the supernatural, is a doctrine or ideology of materialism, the materialist monism of which we spoke in "Sages and Scientists." So much so, the whole enterprise of modern skepticism depends on the viability of materialism as a cogent and compelling belief-system. But, to be intellectually viable, materialism must address and explain the three primordial brute facts of human experience that we've outlined. The first is the existence of intelligent systems that cannot have arisen out of inert matter. The second is our experience of consciousness and thought, and we have seen earlier how this leads ineluctably to the recognition and affirmation of an infinite Intelligence that grounds life, consciousness and intellect. The third brute fact is the human mind's inability to remain at the level of brute facts. We seek explanations for everything and it's the quest for explanation that drives and underlies most human enterprises, including the natural and social sciences. In all three areas, materialism has proven to be helpless and so it's not a rationally viable ideology.

You ask why many thinkers are atheists. On the intellectual level, an awareness of God's existence depends on our grasping certain fundamental insights like the three hard facts I've outlined. There's nothing automatic or inevitable about such insights. While both start with the same set of facts, the atheist merely sees these facts while the theist notices them. If we fail to pay attention to our immediate experience or ignore the various dimensions contained in the data, we can't perceive the transcendent ground of all things and events.

But atheism isn't strictly an intellectual matter. The psychologist Paul Vitz has figuratively put many of the most famous atheists on the couch and come to the conclusion that atheism is a neurosis. Vitz contends that the major barriers to belief in God are not rational but neurotic psychological barriers of which the unbeliever may be unaware. Intellectuals may become atheists to gain social and academic acceptance and for personal convenience. But there are also psychoanalytic motives. Reversing Freud's claim, inherited from Feuerbach, that belief in God is a wish-fulfillment driven by a desire for security, Vitz points out that within the Freudian framework, atheism is actually an illusion caused by the sub-conscious desire to kill the father and replace him with oneself. The well-known skeptic Voltaire vehemently rejected his father, and even refused to take his father's name.

But this explanation is not the whole story. In Vitz's view, it's the defective father hypothesis that covers a wider range of data: a child disappointed in the earthly father finds it impossible to believe in a heavenly Father. As evidence for this thesis, he cites the case histories of various well-known unbelievers: Sigmund Freud himself was deeply disappointed in his father, a weak man; Karl Marx did not respect his father; and the young Ludwig Feuerbach was deeply hurt by his father. The death of a father is sometimes also seen as a betrayal; Jean- Paul Sartre's father died before he was born and both Bertrand Russell and Albert Camus lost their fathers when they were very young. Vitz supports the defective father hypothesis with excerpts from the personal correspondence of these atheists that illustrate the true sources of their rejection of God.

Atheism is essentially a rejection of the Father. And it is only when we overcome the barriers between our fathers and ourselves that we can truly recognize and celebrate the Father who brought all things into being.

> *Geek:* I have to confess that your diagnosis of atheism as a neurosis with roots in childhood conflict strikes a chord with me. At the best of times I've had an uneasy relationship with my father. Now that you mention it, I know I've always felt vaguely resentful around him perhaps because he never appreciated what I achieved from my earliest days. There was a definite barrier between us. But I regret now that I never could bring myself to thank him for all he did for me. What I'm curious about at this point is how this alleged neurosis ties into your concept of God. All that you've said so far about the God-hypothesis has seemed academic and abstract. I can't see where you inject personality and paternity into a cosmic energy-source, so to speak.

Guru: Well, God is not simply a concept or a hypothesis. God is a living reality that is both personal and fatherly inasmuch as he fathered all things out of His infinite love. God is a Mind, not a force, or an energy-source for that matter, a Heart, not an impersonal law.

> *Geek:* All talk of God as personal is allegedly anthropomorphic, God made in the image of humanity, so to speak. Every human person is necessarily imperfect with all the limitations entailed by personality. This doesn't square with the lofty idea of a Supreme Spirit that is infinite and absolute.

> To be perfectly honest, I accept the existence of a kind of cosmic spirit in line with Einstein's idea of a spirit at work in the laws of nature. Even Hume and Darwin accepted the possibility that there is a deity behind the world. But the idea of a personal deity is a radically different notion.

Guru: Let me applaud you on your open-mindedness. You've lived up to your commitment to base your judgments on the evidence you find adequate. I'm not trying to be patronizing, but I think that recognition of a creative force behind the world is a monumental step forward. What this force is like is the obvious next question. I think I've shown you why any Source of all that exists necessarily has to be perfect and without limitation. And you've followed this train of thought to the conclusion that God can't be personal because personality is a limitation.

The premise that there can be no limitation in God is certainly valid. Equally, I agree that human personality is inherently limited and it's senseless to visualize God as some kind of superhuman person.

But here's the rub. Not to be personal is to be less than personal. To be impersonal is to be sub-personal; there's no getting around this. A force doesn't have a mind and is therefore neither intelligent nor free. And clearly God in any coherent sense of the term can't be incapable of intelligence, intention or thought. As for Einstein, I've already told you that he was inconsistent in thinking and talking of God as personal while in theory rejecting the idea of God as personal. Since he didn't have a highly developed theology of God, these implicit contradictions didn't seem to bother him. Perhaps a cogent presentation of God's transcendence and immanence might have given him a consistent and coherent model for thinking about God.

Beyond the twin errors of God as sub-personal and finitely personal, we come to the conclusion that God is infinitely perfect in the essential dimensions of personhood: intellect, will, and freedom. To think, to act, to choose is to be personal, and as infinite Thinker, Actor and Chooser, God, is infinitely personal.

Many scientists have spoken of the Mind of God, and I heartily second most of what they say in this regard. But we cannot forget the Heart of God. If infinite Intelligence is the Mind of God, then infinite Will, the capacity to intend, plan, choose, and act, is the Heart of God. The same laws of nature that show the Mind of God also show the Heart, for these laws are manifestations of both incredible ingenuity and exquisite concern for our life and well-being.

> *Geek:* All I have left are a few nits. I think you've fashioned a compelling vision of God. I don't think there's anything more that can or needs to be said. So what next?

꣠

Chapter

Seeing God Through the Eyes of the World

*G*URU: This is where our journey has brought us: to an infinite source of all being that is Mind and Heart. I'd like to come full circle at this point and return to my initial theme of perceiving God through the wonder of the world.

One way of looking at the hierarchy of the universe is in terms of different kinds of seeing.

1. On the lowest level we have the seeing of frogs, fishes and insects.

2. Second, there is the physical perception of higher animals that involves certain forms of consciousness.

3. Third, the seeing of the mind, which is the ability to perceive meaning.

4. Fourth, the seeing of infinite Intelligence in the workings of the world.

5. Finally, the direct and immediate vision of the divine that is possible only after death.

I would like to suggest that our discussion should now move beyond an academic exercise to this fourth level of seeing, the discovery of God in everyday experience. We can and should experience the activity of infinite

Intelligence at every instant and every level of our being and hear the divine Heartbeat through the processes, events and laws of the world.

But to see clearly our minds must be alert and our hearts must be pure, and this is true most especially of seeing the divine through the eyes of the world. We can know God most clearly from His direct action on us, causing us and maintaining us in being. This begins slowly, then swells into a tidal wave of energy and ecstasy.

But this stage of our journey is one each one of us has to make on our own. I think we've come about as far as the intellect will take us and as far together as it's possible to go without striking out on our own. I don't want to say, "You're on your own" now because, of course, God is with you at all times and everywhere. But it's a question of becoming aware of His presence. Going back to Nature, I think, would be a great help because there are fewer distractions there. More important, as you experience Nature in all its vitality and vibrancy, you become aware of God causing *us*, maintaining us in being.

Please keep me posted on your progress, my friend, and Godspeed!

A Theophany

ear Madvha:

Thanks for the tip. As you suggested, I spent the last few months alone with nature, reflecting on the mystery of existence: the inexplicable appearance of the world, life, consciousness, and the rational self.

And, Madhva, it was in the mystical mountains of Munnar in South India that I came face to face with God, the ever-existent and the all-perfect. On a beautiful day in August, as birds sped past and the mist straggled through the hills, I became suddenly aware that every breath I take, every move I make, every musing of my mind was somehow directly held in being by God. Countless chains of cause-and-effect, innumerable agents and vehicles, all ending in HIM, always and everywhere, here and now. Even as I thought through this, even as the awareness of my dependency sank in, it became clearer than ever that Nature and the world are parts of a Whole, images of the Supremely Real, pathways to transcendent Splendor.

But the very next day I sank into a stupor of darkest depression as I pondered death, oblivion, all those hideous fears of nameless dread that have always plagued me. Yes, all things point to supreme Intelligence, stupendous beauty. But why and wherefore are we? All that I learnt in the last few months seemed to evaporate and vanish like a mirage. I was left only with the terrible thought that I would die and so would all those I hold dear.

But, as they say, the darkest night sometimes precedes the brightest dawn. In the most unexpected and yet the least dramatic possible way, I heard from God. It was not a voice or even a thought. It was, how should I put it, a Father, gentle, tender and infinitely moving. I was aware at one and the same time that He was the Mind who thought a trillion galaxies and the Heart that felt every experience I ever had. Beyond this, I have no comprehensible way in which I can describe the nature of my encounter.

I can simply report to you what I learnt about God from my encounter. To make it easier, I will present it in the first person, as a message to us from the Father of all:

> *I want this whole world to know that there is a God and a Creator. This God is unknown to them; they do not know that I am their Father. I am the ocean of charity. My paternal love extends to all without exception. Nor do I exclude different societies, sects, believers, unbelievers, and the indifferent. I enfold in this love all the rational creatures that make up humanity. How could I leave you alone after having created you and adopted you through my love? I follow you everywhere, and I protect you always, so that everything may become a confirmation of my great liberality towards you, in spite of your forgetfulness about my infinite goodness. This forgetfulness makes you say, "Nature provides us with everything, it makes us live and die." I speak to you through the smallest flower about my beauty and the depth of love with which I created you. I make you a promise: call me by the name of Father, with confidence and love, and you will receive everything from this Father, with love and mercy.*

APPENDIX
A Hundred Wonders of the World

B‍Y WAY OF SUMMARY I propose to lay out here a hundred wonders of the world. For the most part, I have already described these in my main presentations and this appendix serves as a quick reference guide. The hundred wonders are classed under five categories:

1. Microverse,

2. Macroverse,

3. Life,

4. Homo Sapiens and

5. Meta-Terrestrial Intelligence.

In the present context, a wonder is any phenomenon or hard fact that intrigues and awe-inspires.

1. MICROVERSE

1. Fields
Modern physics tells us that the most basic things in the world are fields, energy concentrations that occupy space and follow specific laws. Particles are simply manifestations or "excitations" of fields. At a physical level we're nothing but fields.

2. Vacuums

Vacuums too are energy fields. Random fluctuations convert the energy in a vacuum into pairs of particles and anti-particles. These vacuums are all around us.

3. Electrons and Protons–1

Every electron in the universe has the same charge. Each follows the same laws, orbiting the nucleus and shooting out a photon when it collides with another electron, for instance. Each has a natural life span of ten billion trillion years. A proton is 1836 times bigger than an electron but the two have equal yet opposite charges (proton=positive, electron=negative); the electron to proton mass ratio is the exact proportion required for molecules to form. Stephen Hawking notes that any difference in the charge of the electron would have meant that stars either could not (a) burn hydrogen and helium or (b) explode, both of which were required for life.

4. Electrons and Protons–2

Grains of sand and human brains, says Gerald Schroeder, are made up of the same protons, neutrons and electrons. One mix is passive, the other dynamic, but they're made up of the same components.

5. Quantum Wonders–1

Is a subatomic entity, e.g., a photon, a wave or a particle? It depends on how you measure it. When measured in one way, it acts as a particle; in other cases, it acts as a wave. That's because it originally exists in a simple quantum state without wave or particle properties and acquires these properties only when it's measured.

6. Quantum Wonders–2

It's not possible to know both the momentum and the position of a subatomic particle. In measuring one property you alter the state of the other, so you can only measure either momentum or position precisely.

7. Quantum Wonders–3

Despite the intrinsically unpredictable and incredibly complex nature of the basic building blocks of the world, i.e., the quantum realm, everything surprisingly settles down into orderly, precise and predictable structures at the macro level. How does this transformation take place? We don't know, but it continues to take place all the time. Neither do we know why certain precise and invariant constants, such as Planck's constant, reign in the quantum realm.

8. The Atom

A grain of sand, which has a size of about 100 microns, is made up of a million atoms and each atom is about a millionth of a millimeter. Each atom is a universe in its own right. At its center is the nucleus made up of protons and neutrons which is surrounded by a cloud of electrons. 99.9% of its volume is an empty vacuum. There are 92 different kinds of atoms, and these constitute the periodic table of naturally occurring elements.

2. MACROVERSE

9. The Big Bang–1

Some 13.7 billion years ago, an infinitesimally small quantum state of being billions of times smaller than a hydrogen atom underwent a phenomenally rapid expansion that produced the colossal cosmos we inhabit today.

10. The Big Bang–2

A universe made up of hundreds of billions of stars and galaxies existed in its earliest stage as a subatomic-sized particle! The size of the universe, 10^{-33} cm, and the interval of time in which it existed in this state, 10^{-43} seconds, were the smallest possible multiples beyond which division is not viable. So what laws governed the universe when it was point-sized? The laws of the universe as a whole (General Relativity) or the laws of its tiniest components (Quantum Physics)? We don't have the answer, at least as yet, since there is no unified scientific theory that encompasses the very large and the very small.

11. The Big Bang–3

For the first few hundred thousand years of its life, the universe was nothing but GAS and radiation.

12. The Big Bang—4

All of the elements, chemicals and materials with which we're familiar were born in a cosmic game of "Survivor." Although there was initially an equal amount of quarks and anti-quarks, somehow a tiny surplus of quarks emerged in the particle-anti-particle annihilations of the early universe. For every thirty million anti-quarks, there were thirty million and one quarks. Thus one quark survived every time thirty million quarks and anti-quarks destroyed each other and this incredibly small proportion of surviving quarks constitutes all the matter in the universe. There is no current explanation as to why there was, fortunately, an excess of quarks.

13. The Big Bang— 5

What triggered the expansion of the universe and why does it appear to have the same rate in all directions, and this over a period of billions of years? And how did it come to have a rate that's just right? If it was just slightly faster, stars could not have formed and if it were just a little slower, one part in one hundred thousand million million slower, says Hawking, it would have contracted before there was time for the stars to form.

14. The Size of the Universe

The observable universe, says the astrophysicist Robert Jastrow, is made up of a staggering 100 to 200 billion galaxies. On the average, these galaxies are themselves made up of an average of 100 to 200 billion stars. By any reckoning we live in a universe with trillions of stars like the sun!

15. Fine-tuning—1

The astronomer Martin Rees lists six numbers that made our existence possible, two pertaining to the basic forces, two concerning the structure of the universe and two relating to the properties of space. There is no explanation as to why these numbers obtain. Rees notes the extraordinary fact that there's no dependence of any one of the numbers on the others so that you couldn't deduce any single value from any one of the others.

(Rees #1): The conversion of hydrogen into helium generates the energy fueling the sun. Precisely 0.7% of a hydrogen atom's mass turns into energy in the formation of a helium nucleus. If the percentage had been 0.6%

then there would only be hydrogen atoms in the universe and no stars or elements would form. If it had been 0.8% the bonding would have been too strong and there would be no hydrogen available for energy.

16. Fine-tuning–2
(Rees #2): The strength of the electrical forces holding atoms together is hugely greater than the gravitational force between them; any decrease in the ratio between them would have decreased the life-span of the universe and the inhabitants of the world would not have gotten bigger than insects.

17. Fine-tuning–3
(Rees #3): If gravity had been any stronger than it is, then the universe would have collapsed before life arose; if it had been any weaker, then galaxies and stars couldn't form.

18. Fine-tuning - 4
(Rees #4): The seeds of the future structure of the universe were sown in the Big Bang. There's a ratio of $1/100,000$ between the two fundamental energies that enable the formation of galaxies and planets. A smaller ratio would have produced an inert universe of gas and a larger one would condense matter into black holes so that stars couldn't be born.

19. Fine-tuning–5
(Rees #5): The three observable spatial dimensions of the universe, length, breadth, and height, are the right number from our standpoint because neither two nor four dimensions would permit life.

20. Fine-tuning–6
Over billions of years, helium and hydrogen were cooked at the centers of stars to produce heavier elements like carbon and oxygen. As the stars exploded, these elements flew through space and condensed to form stars and planets. Life was possible only because these elements became available. Moreover, the universe has an optimal combination of stars that are hot energy sources and cold planets, optimal because the chemical reactions required for life would only take place with this particular blend.

21. Fine-tuning–7

(Rees #6): Oxygen constitutes 21% of the earth's atmosphere. If it were slightly less, we would suffocate. A tiny increase, on the other hand, would burn up everything on earth.

22. Fine-tuning–8

Any change in the numerical value of fundamental constants such as the strength of gravity and electromagnetism and the speed of light in a vacuum would be fatal to the formation of life. In addition to the well-known constants, says John Barrow, our existence is dependent on 25 basic constants that govern the masses of elementary particles and their interactions.

23. Space

What is space and what's outside space? The idea that space stretches on and on without any outer boundaries (how can there be any boundaries after all) is mind-boggling. But here we may be seduced by words. *There's no such thing as "space" that exists over and above and alongside all the things in the universe.* All talk about space is talk about relationships between two or more pieces of matter (i.e. energy) just as talk about a stock market simply refers to transactions between two or more individuals. By space we mean simply paths along which it is possible for particles to move or fields to act. The quantity and the distribution of matter in the universe determine the shape and extent of space. This still leaves us with the ever-present question of the origin of matter and its space-time.

24. Time

When did time begin and what time was it before t=0? Einstein, above all, showed us that time is not a flowing river that exists independently of events and agents that cause events.

But agents, such as persons, life-forms, fields, particles, are real, events are real, causes and effects are real, and so time is real. It has been said that space-time *is* the interaction of particle-events. Stephen Hawking notes that he and Roger Penrose proved, within the framework of the mathematical model of General Relativity, that time should have a beginning.

25. Energy–1

Energy is everywhere around us and yet we're barely aware of its presence.

We know that different forms of energy are converted into each other given a conversion mechanism. For instance:

1. The formation of helium nuclei in the sun creates heat energy.

2. This heat is emitted as photons, a form of electromagnetic energy, to our planet.

3. On earth, through photosynthesis, the photons help create chemical energy in plants.

4. This passes on to animals when they consume plants to perform energy-expending activities.

5. Energy from the sun stored as chemical energy in coal or oil is converted back into heat and finally into electricity in a power plant.

We are familiar with such displays of energy as velocity and temperature and pressure and have found that the sum total of mass and energy in any system will remain the same.

26. Energy–2
But where did energy, the energy fields in vacuums, for instance, come from? Martin Rees acknowledges that it is basically mysterious how empty space could have energy associated with it. And we know nothing about the origin of energy or the primal field.

27. E = mc²
The energy in a single gram of any kind of matter, reports Gerald Schroeder, can boil 34 billion grams of water into steam. With his $e = mc^2$ equation, Einstein showed us that matter is concentrated energy: the energy e contained in any material object at rest is derived by multiplying its mass m by c^2 the square of the speed of light, 186,283 miles per second.

28. Gravity
Although Newton and Einstein formulated the laws governing the motion and interaction of matter, we do not know why energy or matter is subject to inertia and gravity. Why do they resist motion or feel gravitational force?

29. Motion–1
One of the great discoveries of modern science is the revelation that every-

404 The Wonder of the World

thing is in motion. For instance, everything on earth is hurtling through space at thousands of miles an hour. If you stand still or sit down, like it or not you will continue to rotate around the sun and the sun itself will be rotating around a galaxy that in turn is racing away from the other galaxies. Moreover the particles that make up your body whirl around ceaselessly, and they continue doing so even when you're dead!

30. Motion–2

So how did this motion originate? We know that the earth moves because of gravitational force: the sun pulls the earth and consequently the earth orbits it. Gravity itself is the result of the curvature of space and space curves because of mass. But why does mass cause the curvature of space? Currently there's no answer to this, and even if a scientific explanation were to emerge, we could then ask why that particular state of affairs obtains. The same is true of motion caused by other physical forces, such as the expansion of the galaxies from the Big Bang, the motion of an electron around the nucleus. In all these cases, we come to a point where we have no further explanation for the way things are, other than to say that's the way they are.

31. Light

Light is arguably the most fascinating form of energy. At every instant of our lives we're surrounded by photons, some dating back to the origin of the universe, speeding by at 186,283 miles per second, which is the speed limit of the universe. In one year a photon travels some six trillion miles. These are the photons that chemically react with our retinal cells constantly to give us sight and with chlorophyll to generate the energy required by plants and us. Light is also today's most potent medium of information transmission: information is transmitted via photons rather than electrons.

32. Electromagnetic radiation

If we think cell phones, radar guns and radio transmitters are works of genius, let's not forget that we're simply hitching rides on the freely available electromagnetic spectrum. James Clerk Maxwell didn't bring about the marvelous unification of electricity, magnetism and optics: he simply discovered it.

33. The Sun
Every second the sun turns 44 million kilograms of mass into energy. Also, sunlight has the precise color required for chlorophyll to absorb it and perform photosynthesis, without which we wouldn't be here to read this.

34. Water
Water, a molecule made up of two atoms of hydrogen and one of oxygen, is the miraculous solvent that serves as the medium of life. It has all the properties that are essential for it to serve as the cradle of life: cohesiveness, temperature storage, solvency, low viscosity, and chemical reactivity. Water makes up 80 percent of the mass of our cells. It is also believed to have played a key role in the formation of stars and solar systems.

35. The Hydrosphere
Michael Denton points out that the hydrosphere keeps a constant level of the 25 elements required for life through interlocking cycles, for instance, the carbon and magnesium cycles. What's especially stunning is the fact that this has continued over a period of four billion years!

36. Snowflakes
We're captivated by the somber silence of snowfall, the majesty of snow-capped mountain peaks. But the microstructure of snow is no less fascinating than its macro-manifestations. Every snow crystal is hexagonal but within this basic six-sided shape there are endless intricate permutations and combinations so that virtually very flake is unique. A collision of atmospheric dust particles and droplets of water gives us a spectacle that is as structurally ingenious as it is esthetically elegant! The three most striking things about snow are its origin, its magnificence and its transience. What is the return on investment on such a colossal production? Is there some underlying beauty about reality that has to constantly manifest itself?

❧

3. LIFE

37. Life–1
On the biological level, life is characterized by such activities as nutrition, metabolism, growth, response to stimulus, self-motion and replication and the complex interplay of nucleic acids and proteins. On a meta-scientific level, life adds a wholly new dimension to the material universe. This is our concern here.

38. Life–2
First, since each and every living being acts or is capable of action it can rightly be called an agent. And since these agents are capable of surviving independently, they are therefore autonomous agents. Every such agent, i.e., every form of life, is fundamentally purpose-driven. That is to say, all of its activities are driven to meet specific goals, whether to nourish itself or reproduce, and the agent does this for itself, on its own.

39. Life–3
Second, every one of these autonomous agents is intelligent because it performs intelligent processing of messages via DNA (see below). David Berlinski has shown that each molecule of DNA is embodied intelligence because it contains information in a set of signs that have their own meaning.

40. Life–4
Third, it seems clear that there is a hierarchy of intelligent agents. At the most basic level you have unicellular organisms, and these include cyanobacteria and other microbes. At the next level you have plants. Tony Trewavas holds that the computational capacity of plants is as good as that of many animals: they acquire information, plan outcomes and use "complex molecular signaling pathways." Beyond plants there are animals and within the animal world there is, of course, a further hierarchy, from the capacity for locomotion and reproduction to the operation of the senses, instinct, memory. At the top of the ladder of life in the world is the human person who is self-conscious, rational, and free.

41. Life–5
It seems to me impossible that autonomous, intelligent agents could come

to be in a universe of undifferentiated matter. The oldest fossils we have of life, going back over 3.5 billion years (some experts say that the earliest definitive samples are 1.9 billion years old), is of modern photosynthetic cyanobacteria that eat, excrete, metabolize, move and reproduce. These bacteria meet the criteria of intelligent agency: they are source and cause of their actions with goal-driven, inbuilt dynamisms of self-generation and information-processing. The basic template of life, with all the enormous complexity of DNA and RNA, came fully formed billions of years ago with the first appearance of life.

42. DNA-1

DNA is the biological language of life, and its letters are Guanine, Adenine, Cytosine, and Thymine. It is a data repository of genetic information that transmits hereditary characteristics to future generations. Copies of the information encoded in DNA are transmitted via another nucleic acid called RNA to the assembly line where proteins are manufactured. DNA, RNA and proteins store information, communicate, construct, synthesize, repair, replicate, and do all of this through systematic, cooperative, collaborative interaction.

43. DNA-2

The DNA present in each cell contains three and a half billion nucleotide bases, is about two meters in length and has more information than three complete sets of the *Encyclopedia Britannica*. With its four-letter language, which allows the formation of a virtually infinite number of genetic combinatorial sequences, DNA spells out the genetic code of every species.

According to Gerald Schroeder, chemical laws may explain the bases, sugars and phosphates of DNA but not its information content or the intelligence driving it.

44. Protein Folding-1

All proteins in living beings are made from different sequences of just twenty organic molecules called amino acids. Proteins have the extraordinary ability to assemble themselves without external intervention. This self-assembly is a process called protein folding whereby a given sequence of amino acids forms a specific three-dimensional structure: it becomes a particular protein with a precise structural or functional role.

45. Protein Folding–2

Every cell in the body other than sex and blood cells, says Gerald Schroeder, makes two thousand proteins every second from hundreds of amino acids. This process is so complex that, as reported in *Scientific American*, a supercomputer, programmed with the rules for protein folding, would take 10^{127} years to generate the final folded form of a protein with just 100 amino acids. *But what takes a supercomputer billions of years, takes seconds for real proteins.*

46. The Cell–1

Each cell contains 10^{12} bits of information and any one cell has coded within it the information required to build a copy of the organism's whole body. Every cell is a high-tech factory with complex transportation and distribution systems and huge libraries of information.

47. The Cell–2

E.J. Ambrose has classified the intricate structures within a cell under five heads:

1. The power house whereby the cell receives and regulates a continuous flow of energy,

2. The machine tools—proteins are the tools that run all the activity in the factory,

3. Raw materials—elements like nitrogen and sodium that are passed through various production lines till the required end-product has been generated,

4. The factory gates—the factory is protected by a gate of sugar molecules and a gate keeper made up of fat molecules for letting in raw materials,

5. The fertilizer plant—the factory requires ammonia or nitrates to function.

48. The Cell–3

Our immune and nervous systems and our anatomy as a whole are able to function because our cells communicate with each other more accurately than users of any telephone network. Messenger molecules, e.g., hormones, from one cell dock with the receptor molecules of another cell. The messages are then sent through signaling pathways and acted on by molecules in the recipient cell.

49. A Field of Grass

A field of grass may be thought of as a bank of sophisticated computers processing vast quantities of information. In fact it's far more powerful than any of today's computer systems because it responds intelligently and instantly to a whole host of sensory inputs from gravity to sunshine to an array of chemicals.

50. Flowering Plants

There's something magical about flowers: morning glories, roses and lilies, daffodils and dahlias, magnolias and marigolds, petunias and poppies, primroses, daisies, snapdragons. Flowering plants are believed to have appeared 130 million years ago and now include some 235,000 species. Almost all our non-animal foods come from flowering plants. The flower with its stamen, petals, pollen, and stigma is the exquisite reproductive machinery of the plant. We know the origin, structure and function of these beautiful beings, but the central issue remains on the table: how did the laws that work this magic come about? We can touch and smell and see "flower power" here and now. But we are left to wonder who or what wrote the code that gives us the pinkness of a rose and the intensity of its fragrance. Is it someone or something that "says it" with flowers?

51. Insect Flight

How do insects fly and hover? Initially it might seem that the aerodynamics involved would work against flight. For instance, how is it possible for a bumblebee to fly given that its wings are too small to support the lift required by its weight? Moreover, insects, unlike airplanes, continually flap their wings—and this is hard to square with theoretical calculations. Michael Dickinson points out that fruit flies, which know nothing of aerodynamics, nevertheless utilize vortex production, delayed stall, rotational circulation and wake capture as they effortlessly stay aloft while flapping their wings about 200 times a second.

52. Songbirds - 1

Who came up with the idea of songbirds: larks and wrens, mockingbirds and starlings, finches and wood warblers? Fossils that actually look like birds date from the Cretaceous period. By the Tertiary period, about 60 million years ago, various modern birds from waterbirds to songbirds turn up. In all, some 10,000 species of birds are found on earth today. It's believed that ecological factors largely caused the diversity in birds. While these and other avian theories are useful, it leaves the primary question entirely unanswered. And that's how matter allowed all this to happen. Ecological factors may have unlocked the capabilities of the organisms in question, but where did these innate capabilities, the unique flight feathers with its aerodynamic capabilities, the respiratory system found in no other mammal, come from? And how is it that such adaptation to ecology was possible? Take it down to the genetic level; even with random mutations, the genetic material must ultimately have contained the capabilities we now see manifested in sparrows and robins, meadowlarks and woodpeckers. Or if you go back before the origin of life, you ask how the matter fields existing before the origin of life acquired all the capabilities that were to manifest themselves after the genesis of life and living beings. Were those innate capabilities, and if so, how did matter come to have such properties and how is it then able to manifest them?

The point is that there's something involved here beyond quantum fields, and this right from the very beginning. When we see birds take off with a runway of inches or fly across continents for months at a time, we have a right and an obligation to be wonder-struck. To lightly dismiss them as products of evolutionary adaptation is to ignore the questions of (a) where did the entity that does the evolving come from, (b) how did it come to have its innate capabilities, and (c) why is reality structured so that adaptation is possible.

53. Songbirds—2

So the next time you see a bird, don't view it simply as a twig on the evolutionary tree, a vehicle of natural selection or matter in motion. Consider its displays of intelligent, even extraordinary, behavior, and assess these displays on their own terms. Listen to it sing. This is not just a disturbance of fields (although it is that) or phonons gone berserk. It's a bird singing. And it's here, now. Its existence is as marvelous and moving as the most beautiful painting. Yes, it emerged through progressive modifications from primordial times but the endgame is what matters, as surely as the end painting makes sense of the initial pigments. Its properties of intelligence and beauty bespeak a source, in the ultimate analysis, that is intelligent and beautiful. Like the painting it most certainly is a manifestation of matter fields, and like the painting the form taken by the manifestation cannot be explained without reference to mind.

54. Dogs

Dogs! Faithful companions, selfless servers, tail-wagging, lick-happy bundles of joy. Labradors, poodles, basset hounds, alsatians, bull terriers. How did we wind up with our oldest and "best friends?" The fossil evidence of this moving friendship of the species goes back over 14,000 years. How dogs and humans first took to each other is anyone's guess although several plausible theories have been advanced (they were good co-hunters with a remarkable sense of smell as well as reliable guards). Dogs were domesticated by the ancient Egyptians and Greeks. With their 78 chromosomes, they have been bred in all shapes and sizes. Dogs serve us in numerous capacities: as alarm systems, "seeing eyes," rescuers and simply as companions. Panting, wagging, slobbering, leaping, fetching, peering deep into our hearts—what would life be like without these kindly friends?

55. Reproduction

Neo-Darwinists like to say that there's no purpose in nature. But neo-Darwinists themselves have to assume capabilities of self-reproduction at the earliest stages of life. Yet reproduction is an irreducibly purpose-driven act, one that can't simply spring from matter. How is it that the first living beings had the powers of replication? How is it that life came with this fundamentally purposive capability pre-installed? John Maddox, the former editor of *Nature*, admits that we don't know how sexual reproduction itself evolved despite decades of speculation. Replication is the engine that runs evolutionary theory. *It's the horse of reproduction that draws the cart of natural selection.* If you put the cart before the horse, you won't get started. Who came up with the idea of replication, and who then imprinted material structures with a vast variety of replicational capabilities?

56. The Cambrian Explosion

What triggered off the Cambrian Explosion? All the body plans of all living animals were formed in a period of 5-10 million years, a relatively short time in the multi-billion year history of life. If it was a question of certain genes being switched on, why did it happen at that particular time and why all of a sudden and why never again?

57. Genes

John Maddox comments that virtually nothing is known of the following:

- ♦ How genes switch on the suite of genes hierarchically beneath them
- ♦ How genes give cells their particular nature
- ♦ How the complex system of genes came to be.

58. The Tree of Life

No matter how much the randomist wants to say that there's no progress or direction, the fossil record is a record of progression from unicellular to multicellular life, from amoeba to *Homo sapiens*. Whether this end-point was reached accidentally and purposelessly is a question that science can't answer because science as science can't discern purpose or goals. You can certainly interpret the data as indicating no purpose. But it seems extraordinary that:

1. There is a clear progression,
2. There are exact matches between sudden leaps forward in the tree of life and the existence of environments that allow these upgrades to survive and thrive,
3. The matches were made in relatively narrow windows of time.

59. The Evolutionary Story–1

Often enough, many of the wonders of the world are dismissed as inevitable products of evolutionary processes or simple instances of natural selection at work. But these responses miss the point altogether: natural selection can't work if there were no laws of chemistry that are themselves dependent on particular laws of physics. What requires explanation, and can't be explained by evolution, is the existence of these finely tuned laws.

60. The Evolutionary Story–2

Evolution calls for adaptation to environment, but where did this ability to adapt come from? And what is the source of the "raw material" that does the adaptation and of the environment in which it adapts?

4. HOMO SAPIENS

61. The Five Senses

Perhaps our senses can bring us to our senses! In seeing, hearing, smelling, tasting and touching, a mechanical stimulus is transformed into a nerve signal that is sent to the brain and then converted into a conscious state. Despite decades of scientific study spent in understanding the network of proteins, ions, signals and cellular structures involved, the bridge between these two worlds, the external stimulus and the corresponding sense-perception, remains as much a mystery today as ever. On the one hand, we have an efficient chain of precise physical processes that monitor, transmit and respond to an immense variety of sensory inputs. On the other, we have a mysterious and radical conversion: the merely physical becomes something of which we are conscious, something in which we participate.

62. The Sound of Music

The vibration of something physical like the vocal cords produces sound. It is propagated by the displacement of particles like air molecules and picked up by complex anatomical structures called ears. But how can a mere wave of mechanical energy become a package of meaning (language) or a thing of beauty (music)? How can a vibration in the air, a fluctuation of pressure serve as the foundation for a haunting melody that opens windows into another world? If the goal is the creation of sound then a most improbable collection of energies, particles and organs has been assembled to attain it. Which leads us to ask: is our medium of communication itself a communication? Is the packaging of intelligence itself intelligently packaged?

63. Seeing–1

The retina, which is about one centimeter square and one half millimeter thick, has a 100 million neurons that are functionally divided into five classes: photoreceptor, bipolar, ganglion, horizontal and amacrine cells. As it carries out its image analysis, retinal neurons communicate with each other through synapses using chemical messengers called neurotransmitters. To distinguish colors, the retina uses the photoreceptors, which are made up of millions of rod and cone cells. The rods give the intensity of light distinguishing between light and dark while the cones read its wavelengths.

64. Seeing–2

A chain of chemical reactions is triggered off when a retinal cell absorbs light. A single photon acting on the rhodopsin protein in a rod cell sets off cascades of enzymatic activities that translate this information into a signal that can be processed by the nervous system. The retina processes ten one-million-point images every second. Through chemical amplifiers it produces up to 100,000 messenger molecules from a single photon. To match the retina's processing power, a robot vision program would have to perform 1,000 million computer instructions per second. It has also been estimated that computer simulation of the processing performed by just one retinal nerve cell in one hundredth of a second would call for the solution of 500 simultaneous non-linear differential equations 100 times.

65. Seeing–3

At one time it was thought that the eye evolved separately in different life-forms. Today, as Stephen Jay Gould has said, there is no longer any separate story of evolutionary origins for the eye because an underlying master-control gene, common to all phyla, positioned the fundamental structure of the eye on one and the same genetic pathway. Fully functional eyes appear right at the beginning of the Cambrian explosion. The Pax-6 gene governs the development of this delicate and intricately organized organ in the most varied species. A single genetic blueprint was pre-programmed into the very stuff of being.

66. Seeing–4

Why is reality structured so that there can be a symbiosis between light and the eye, between photons and neurons, that results in sight? Light waves, incidentally, are the only kind of electromagnetic radiation with the energy level required for biological systems to detect them and the wavelength of light is just the right size for the high-resolution camera-type vertebrate eye.

67. Seeing–5

To see means to be aware. Awareness is a new kind of reality that transcends the physical and is irreducible to anything else. The objects in our visual field in a certain sense become a part of us, not physically of course but mentally. Now none of this could happen if we did not have our mysterious organ of vision and its equally mysterious interplay with light, but the end-result we experience as visual awareness is something that can neither be predicted from the physical components that make it possible nor be described in physical terms. It's this dimension of seeing, its introduction of the entirely new reality of awareness, that makes it a marvel and a mystery of the first order.

68. The Brain
The brain has a 100 billion neurons that use dendrons and axons to send information across one quadrillion synapses. The 100 billion neurons do 100 million MIPS, i.e., millions of instructions per second.

69. Consciousness–1
We're conscious and aware that we are conscious; we perceive, conceive, remember, imagine, sense, feel, plan, intend, and choose. Not only are we conscious of being conscious, but we're just as clearly conscious that our consciousness is dramatically different from anything material or physical; it has no size or shape, color or smell.

70. Consciousness–2
We have pinpointed the kinds of information processing carried out by the nervous system, and we have hypotheses about what brain mechanisms enable consciousness to access such information. But, says Steven Pinker, we have no idea where sentience, i.e., *what consciousness feels like on the inside*, came from. How can a purely physical universe give rise at random to something that we experience intentionally as qualitatively non-physical?

71. Thought–1
The mind's capacity to grasp meaning is the exact counterpart on the conceptual plane of what takes place during the act of seeing at the perceptual level. When we "see the point" or "see something to be the case" or "know what you mean" or "realize" or "understand" or "comprehend" or "visualize," then we are performing acts that cannot be described or explained in physical terms. All of these acts presuppose the mind coming to grips with something. Meaning is all about reasons and not causes. If I say a poem is beautiful, neither my message nor its truth is reducible to neuronal excitement in given regions of the cerebral cortex. It's all about concepts, reasons and meanings, and not causes and effects.

72. Thought–2

The activity of the mind is especially apparent in the classification of our experience under universal concepts: the mind works on the data of our immediate experience to see what's similar between different objects and to separate these similar elements into universal concepts, e.g. the concept of a dog from our experience of Rover. Conceptual thought is a non-material activity because concepts (e.g., blueness or justice) are not necessarily linked to something physical.

73. Thought–3

Brain and mental events have entirely different properties. Moreover, mental events cause brain events. Thoughts drive the corresponding neural transactions and not the other way round. There's no chicken or egg question here. The thought comes first and, as a result, causes certain brain events. But where did thoughts come from and how? What causes the mental events that cause the brain events? Other brain events? Then, is the thought "our solar system has nine planets" simply a set of brain events caused entirely by other physical transactions in the brain? Clearly it's the mind acting on the brain that is the source of all our experience.

74. Thought–4

How is our thought different from the information processing of a computer? The computer is not conscious or aware of itself doing its processing as we are when we think. Rule-governed activities, such as chess, can be formalized and simulated in a computer program. But remember that it's we humans who do the programming and create and input the appropriate symbols. The computer has no clue about the meaning of the symbols it processes. All that it does is generate electrical pulses. It doesn't know what it's doing. There is, in fact, no "it" to know because it's simply an assemblage of mechanical systems. There is no knowing going on within it; there's simply a continuous flow of electrical pulses through its circuitry, assuming, of course, that it's connected to a power source and appropriately programmed by a user. Real thinking, on the other hand, involves our knowing what we're thinking about, being aware that we're thinking, recognizing that we're arguing or reaching a conclusion and the like.

75. Language–1

The most obvious everyday instance of conceptual thought is language.

Syntactical language is unique to human beings, found even in ancient civilizations and instinctively mastered by children at a very young age. Language is built around the ability to understand. There's no organ, no part of the brain that performs understanding. Words are symbols or codes signifying something, and the coding and decoding activities required for using language presuppose an entity that can endow and perceive meaning in symbols. Can a material object perceive meaning? By its very nature, the act of comprehending the meaning of something is non-physical. And it's something we do all the time.

76. Language–2
Richard Dawkins points out that nobody knows how language began since (a) there's no syntax in non-human animals and it's hard to imagine evolutionary forerunners of it and (b) the origin of semantics, of words and their meanings, is equally obscure. Noam Chomsky notes that we can't derive human language from animal communication systems since our languages are built around syntactical rules.

77. The Self–1
What is it that perceives and conceives, feels and thinks, judges and chooses? It is the self, the center of our consciousness, the unitary unifier of our experiences, that which gives us the identity of being the same person throughout our lives although the physical components of our bodies change constantly. We're not conscious of the self separately from its acts but we're conscious of it as the ground that pervades and unifies our acts, the entity that thinks, wills and feels.

78. The Self–2
The "I" is not present in any region of the brain. Steven Pinker argues that the "I" is not a combination of body parts or brain states or bits of information, "but a unity of selfness over time, a single locus that is nowhere in particular." Now someone might say this "I" is a conglomerate of cells but then I ask what's the center, the hub of the conglomerate. You say there's no such center or that certain brain cells function in this role. There are a number of mini-agents, your cells, but no overall super-agent that runs the show. At this stage, I remind *you* that *you* constantly experience the clear, indubitable awareness of being an "I," of being not just conscious but self-conscious. Of course you could say that your sense of "I," your self-

consciousness, is just an illusion. But an illusion has to be experienced *by someone* to be an illusion, which brings us back to the existence of a center that is conscious and rational.

79. Free Will and Intention

Undeniably, we think we make free choices all the time, whether we decide to get out of bed or choose between two candidates in an election. Likewise, we're aware of pursuing goals, of intending to act in certain ways and then carrying out the intention. The determinist says that all our choices and intentions were pre-determined entirely by prior physical states of the universe. But this claim is both self-contradictory and false to our experience. If determinism is true, only physical causes exist and not reasons or truths, and there is therefore no reason to believe it to be true. The fact of the matter is that our actions can quite clearly be explained in terms of reasons and purposes, not simply in terms of our brain states. There are choices and decisions we make after weighing attractive alternatives, and we make right and wrong decisions sometimes "against our wills." If we have no free will, murderers and embezzlers can't be punished and heroes and altruists can't be praised because all of their actions were caused by purely physical factors over which they had no control. Only in recognizing a trans-physical center of all our willing can we make sense of our experience or have any basis for responsibility. Intention too can't be a purely physical process since the very act of intending something implies conceptual thinking and a deliberation that is performed by a conscious self.

80. The Origin of the Mind

The most obvious thing of all is this: we exist in two worlds. The first is the world of matter, of fields and their manifestation as particles, stars and organisms, and the second is the world of the mind, of conscious acts and rational judgments, intentions and decisions. Both worlds are radically different from each other but they constantly interact. We cannot explain either one in terms of the other. It's simply ridiculous to suppose that feelings and cogitations are simply photons and electrons. And it's just as incoherent to assert that a universe of pure matter with no purpose, no intellect, no consciousness, and no will whatsoever can give rise to conscious, thinking, willing agents. It's almost as if we were to say that given an infinite amount of time, a pen and a paper, without any external intervention, would somehow give rise to the concepts embodied in the

Gettysburg Address. We're not speaking of the words that constitute the speech but the concepts of liberty and equality and justice represented by the words. These concepts as such are so radically different in nature from the physical objects used to represent them that it's simply nonsensical to suppose that the latter could produce the former.

5. META-TERRESTRIAL INTELLIGENCE

81. How did things come to be?
The mystery that anything **is** is a mystery beyond belief. Every time we realize that we or anything around us exists, we are moved to wonder how this was and is possible and how it all began. We're not asking about the physical origin of the universe but about the fact that anything at all exists. Why is there something rather than nothing, and why this particular something?

82. Science and the quest for explanation–1
All of science is built around the belief that there is an explanation for everything. Not only do scientists believe that everything can be explained but they seek to show that the explanation conforms to the laws of nature. None of them say that the universe or anything in it happens causelessly. The quantum theorists among them might hold that events at the earliest stages were not predictable or determined, but they don't deny that there were certain conditions that had to exist, i.e. a vacuum with a precise structure, for certain effects to be produced. So how do we explain the fact that facts can be explained?

83. Science and the quest for explanation–2
Scientists not only believe that everything in the world can be explained but that we can know things about the world. This mysterious correspondence between mind and world is taken for granted by most of us. But Einstein said that the most incomprehensible thing about the world is the fact that it's comprehensible.

84. How does the electron know what to do?
Every one of the billions of electrons in the universe has the same proper-

ties and follows the same behavior patterns. But how were these laws implanted and how are they enforced?

85. The laws of nature–1

From atoms and cells to galaxies and ecosystems, we see intricately ordered activities and operations manifesting carefully defined laws; the laws that affect a quark affect a galaxy. How utterly extraordinary it is that the physical constituents of this world obey any laws at all. Why on earth should stones and feathers and stars follow uniform laws of motion? How can mass-energy be made to conform to the law of its conservation? What tells the molecules of a gas that the product of their volume and pressure should be proportional to their absolute temperature? How do nucleic acids like DNA and proteins know that they can and *should* communicate and construct, repair and replicate? How do genes know when to switch on and why do they continue to do so at just the right time? What compels organisms to adapt to their environment? Why should these and all the other particles and forces, galaxies and gas clouds, cells and chemicals that make up our world follow instructions? Do they know they're supposed to act in that way? Were they programmed to obey orders, march in step? Trillions of quarks and cells, billions of galaxies, all of them doing what they're told to. But by whom and how?

86. The laws of nature–2

Do the laws of nature exist independently of the fields that manifest themselves as particles and forces? If so how do these entities know that they should follow the laws? Or are the laws programmed in them? But where do the programs reside? No matter how much we open up the fields in question we can't find any programs in them. How do gluons know that they should be carriers of the strong nuclear force that bind quarks? How is it possible that a material thing can follow any law whatsoever? We humans follow laws we set up because we're conscious agents capable of intentional action. But how can an inanimate thing be made to do anything, let alone follow a law?

87. Smart universe–1

One great contribution of modern science to our understanding of the world is the revelation that intelligence is all-pervasive. From the forces and particles of the subatomic realm that operate within a framework of precise symmetries to the world-wide web of information-rich process-flows and just-right physical constants to the inexhaustibly resourceful DNA that builds the living world, we live in a "smart" universe.

88. Smart universe–2

The laws of nature are not simply brute facts but constitute a platform of parameters driven toward life. We are struck not simply by the existence of rational principles underlying the physical world but also by the fact that they operate within certain specific ranges to enable the existence of conscious, rational life. The world is rationally ordered, and moreover, it is ordered towards the production of rational life. The laws themselves, it may be said, are intelligent and purposive.

89. The IQ of the universe

Like the sudden appearance and outward expansion of the universe itself, the genesis and growth of life on earth are unmistakable manifestations of extraordinary intelligence, beauty and power. There is a steady progression of foundational transformations that proceed in this sequence:

♦ An initial matrix of unutterably intelligent laws,

♦ The application of these laws in precisely the direction that maximizes the IQ of the whole system so to speak, despite the most inhospitable conditions imaginable from beginning to end,

♦ The manifestation of a hierarchy of radically different kinds of intelligence: matter, life, innately purposive structures, consciousness, and mind.

What's striking about this whole dynamic is the incredible complexity that exists from the start and the systematic appearance of new kinds of things such as life and consciousness that cannot be put together from whatever existed before.

90. Mathematics and the Real World–1

Nothing manifests the existence of intelligence in the world more clearly than the correspondence between abstract mathematical theories and real-world applications. All the laws of physics are mathematical, and yet they are effective at every level of the physical world. Einstein's General Theory of Relativity was a work of abstract mathematics; nevertheless, it was precisely confirmed in experiments years after it was first proposed. Eugene Wigner rightly spoke of the unreasonable effectiveness of mathematics in the natural sciences.

91. Mathematics and the Real World–2

Many scientists today see scientific laws as algorithms for processing information and physical systems as computational systems. David Deutsch has remarked that the universality of computation is the most profound thing in the universe.

92. Mathematics and the Real World–3

The fact that mathematics works so well when applied to the physical world demands an explanation. Why do the laws of nature permit physical models for the laws of mathematics? Paul Davies concludes that the natural world is not just a concoction of entities and forces but an ingenious and unified mathematical scheme. The legendary quantum physicist Paul Dirac remarked in *Scientific American*, "God is a mathematician of a very high order and He used advanced mathematics in constructing the universe."

93. Everything is information–1

For many scientists, the codes and blueprints latent in the workings of the world are best described as information. All the information on a particle, for instance, is represented by the wavefunction that describes its behavior. The quantum physicist John Archibald Wheeler has said that his lifetime in physics was divided into three periods: "everything is particles;" "everything is fields;" and, now, "everything is information." Gerald Schroeder observes that Wheeler "sees the world as the 'it' (the tangible item) that came from a 'bit' (eight of which comprise a byte of information)." Information *precedes* its manifestation in matter.

94. Everything is information–2

Unmistakably, matter, mass/energy in this context, is the primary vehicle of information in the world. Whether it's information programmed by us, e.g. software, movies, books, or communicated by mysterious instruction manuals like DNA, or simply inbuilt as with anything that follows the laws of nature, everything in the universe is controlled by coded information.

But matter is purely a vehicle. How did it become a vehicle for codes and blueprints? We know it takes intelligence to decode the information transmitted by matter. But if decoding requires intelligence, how about the encoding? If information exists prior to matter, what is its source?

95. The Invention of Nature–1

Someone invented everything around us in the human world: cars, cookies, restaurants, microchips, and clothes. All these things were first ideas in the minds of their inventors before they were brought into being. When we turn to nature, we ask who or what thought up photons and suns, DNA and dinosaurs, mass and charge? Who or what is powering the whole enterprise, keeping it always "online?"

96. The Invention of Nature–2

Nature with its systems and laws embodies ingenuity, uniqueness, and plenitude. To the extent that it embodies an Idea, it too is an invention. Of course we can give the physical, chemical and, where relevant, biological conditions that gave rise to every one of the things in nature. And, of course, we can delve into their subatomic antecedents so as to thoroughly chronicle and predict their wavefunctions and world-lines. And we can further classify them under different categories: phylum, genus and species; infrared, ultraviolet, gamma and X ray radiation; quarks and leptons; and the like. And to terminate further inquiry we can take final refuge in the cosmic cauldron that was the Big Bang, the primordial ooze of prebiotic evolution and the fiery hand of natural selection. But the fundamental dimension we are concerned with here goes beyond any description of the development of a thing and its classification under some scheme. We are concerned with the blueprint of its being. On the scientific account, there is a blueprint underlying the existence of all the things that constitute the world. And it's an inventor's blueprint because each thing with its particular conjunction of properties and its ability to follow certain specific laws is framed in terms of an idea and it is an idea that comes to life. Every petal, every quark, every molecule of gas was thought up before it was materialized.

97. Who Holds the Patent on the Genetic Code?

Inventions embody ingenuity. The systems in nature and the laws of nature are far more complex, intricate and innovative than the internal combustion engine or the submarine. In fact, human inventions are impossible without the existence of natural laws and systems. If a set of primordial states and initial conditions gradually unfolded to produce a given thing or being, these states/conditions were programmed at the start to produce this end-result. *Who or what wrote these programs? Who holds the patent on the process and its products?* If we ask why grass is green, our question is not about chlorophyll or molecular structure or energy sources. Given that certain ultimate physical processes gave rise to a blade of grass then its greenness and other properties were contained in those processes from the very beginning.

98. Intelligence is a Hard Fact

The mistake of the materialist is to assume that intelligence is something observable in the same sense and at the same level as molecules and fields and to conclude that it's non-existent if non-observable. Everyone agrees that the activities of quantum fields cause changes that trickle up to the macroscopic level. But how is it that there is such a relationship between subatomic and atomic realms? How is it that there are different levels in the structure of material beings? How is it that there are different properties at each succeeding level? How is it that new properties start appearing in the world? Intelligence is a hard fact of living things and Nature as a whole, that is, if you think the laws of nature, the purposiveness of replication and the grandeur of energy, to name just three examples, count for anything. When you see a painting, by all means admire the polymers and pigments, but don't remain at the ground level. It's like saying that *Macbeth* is really just black marks on white paper or a collage of costumes, sets and people saying certain words from memory. This is not to see what the play really is. It is to entirely miss the play.

99. A Universe Created by Scientists

Imagine this scenario: we assemble the world's most brilliant scientists, living and dead, Einstein and Edison, Maxwell and Marconi, Newton and von Neumann. Is it conceivable that, given any amount of time, they would be able to create a universe like ours with all its precise parameters and laws? First, they would have to generate a universe-building recipe, then find the ingredients required and, finally, cook the dish. No one seriously believes that it's even remotely possible for the best and the brightest to pull off this job. There's also the added complication that they would have to be in existence before they can create everything, themselves included. If we can't imagine the greatest human intellects bringing the universe with its incredible intelligence into being, why would we assign the role to a vacuum (but who brought that into being?) or chance (which is the absence of all intelligence)?

100. The Scientific Journey from the Wonder of the World to the Mind of God

All the processes in the universe bear the earmarks of mind: the mathematical codes employed everywhere in nature; the immeasurably intricate, precisely structured, comprehensively coherent patterns of interaction between energy fields and galaxies, proteins and nucleic acids; the existence of autonomous agents and purposive behavior; and the reality of conceptual thought. All of these phenomena are manifestations of infinite Intelligence. It's a striking fact of history that Copernicus, Galileo and Kepler, Newton, Faraday and Maxwell, Einstein, Planck and Heisenberg, all believed in a divine Mind behind the world and Rationality at the foundation of reality. The answer to the question of why the universe exists, said Stephen Hawking, would reveal to us "the mind of God." And Einstein declared that anyone seriously engaged in the pursuit of science becomes convinced that the laws of nature manifest the existence of a spirit vastly superior to that of human persons.

ॐ

REFERENCES

References

Prologue

"The All-Maker" in *The Rig Veda,* transl. Wendy Doniger O'Flaherty (India: Penguin, 1994), 36.

Rudolf Otto, *The Idea of the Holy,* transl. John W. Harvey (Oxford: Oxford University Press, 1958).

Albert Einstein, *Out of My Later Years* (New York: Philosophical Library, 1950), 58.

Seven Wonders

Albert Einstein, "The World As I See It" in *Living Philosophies* (New York: Simon Schuster, 1931), 3-7.

1. How I Wonder That You Are

Ludwig Wittgenstein, *The Tractatus-Logico-Philosophicus* (Atlantic Highland, N.J.: Humanities Press Inc., 1961), 6.44.

2. How Does the Electron Know What to Do?

Michael Denton, *Nature's Destiny* (New York: The Free Press, 1998), 19ff., 89.

Gerald Schroeder, *Genesis and the Big Bang* (New York: Bantam, 1990), 180.

Gregory Benford, "Leaping the Abyss: Stephen Hawking on black holes, unified field theory and Marilyn Monroe." *Reason* 4.02, April 2002, 29.

Stephen Hawking, *A Brief History of Time,* (New York: Bantam, 1988), 175.

Albert Einstein, *Ideas and Opinions,* transl. Sonja Bargmann (New York: Dell Publishing Company, 1973), 255.

Richard Swinburne, *Is There a God?* (Oxford: Oxford University Press, 1997).

3. Smart Universe

Four Kinds of Intelligence

David Deutsch, cited in Kevin Kelly, "God is the Machine." *Wired*, December 2002.

Albert Einstein, *Ideas and Opinions,* transl. Sonja Bargmann (New York: Dell Publishing Company, 1973), 49.

E.P. Wigner, "The Unreasonable Effectiveness of Mathematics in the Natural Sciences." *Communications on Pure and Applied Mathematics* 13 (1960), 1-14.

Roger Penrose, *The Emperor's New Mind* (Oxford: Oxford University Press, 1989), 430.

John Archibald Wheeler, "Information, Physics, Quantum: The Search for Links" in Wojciech H. Zurek ed. *Complexity, Entropy and the Physics of Information* (Reading, Mass: Addison-Wesley, 1990), 5.

Gerald Schroeder, "With wisdom God created the heavens and the earth—The universe as the symbol of a thought."

4. The Enigma of Energy

Martin Rees cited in "What Came Before Creation?" *U.S. News & World Report* special edition on Mysteries of Science, 2002.

Rupert Sheldrake, "Science and Spirit," *Noetic Sciences Review*, Spring 1987, 3.

Fields

J.R. Lucas, "The Nature of Things." Presidential address to the British Society for the Philosophy of Science, June 7, 1993.

John Archibald Wheeler and Albert Einstein cited in K.C. Cole, *The Hole in the Universe: How Scientists Peered Over the Edge of Emptiness and Found Everything* (New York: Harcourt, 2001), 74.

David Malament cited in Hans Halvorson and Rob Clifton, "No Place for Particles in Relativistic Quantum Theories?" *Philosophy of Science*, March 2002, 23.

Henning Genz, *Nothingness: The Science of Empty Space,* transl. Karen Heusch (Reading, MA: Perseus Press, 1999), 207.

5. The Wonder That is Life

Intelligent Message Processing

E.J. Ambrose, *The Mirror of Creation* (Edinburgh: Scottish Academic Press, 1990), 88ff.

Gerald Schroeder, *The Hidden Face of God: How Science Reveals the Ultimate Truth* (New York: The Free Press, 2001), 189 ff.

John L. Casti, "Confronting Science's Logical Limits." *Scientific American*, October 1996, 103.

David Berlinksi, *The Advent of the Algorithm* (New York: Harcourt, 2000).

Gerald Schroeder, "With wisdom God created the heavens and the earth—The universe as the symbol of a thought."

Intelligence in Nature

Storrs Olson cited in D.K. Parsell, "'Feathered' Fossil Bolsters Changing Image of Dinosaurs." National Geographic News, *April 25, 2001.*

The Origin of Life

John Maddox, **What Remains to be Discovered** (New York: Touchstone, 1998), 369-370.

Werner Arber, "The Existence of a Creator Represents a Satisfactory Solution" in Henry Margenau and Roy Abraham Varghese ed. **Cosmos, Bios, Theos** (La Salle: Open Court, 1992), 142.

6. Homo "Sapiens"

Richard Dawkins and Steven Pinker, "Is Science Killing the Soul?" (The Guardian-Dillons Debate) *Edge* 53, April 8, 1999.

Ian Tattersall, "The Monkey in the Mirror: Essays on the Science of What Makes Us Human," *Scientific American*, December 2001, 59.

Language

Philip E. Ross, "Hard Words." *Scientific American*, April 1991, 138ff
Colin Renfrew, "World Linguistic Diversity." *Scientific American*, January 1994, 116ff.
Ian Tattersall, "The Monkey in the Mirror," 58.
David Braine, *The Human Person: Animal and Spirit* (Duckworth, London, 1993).

The "I"

David H. Lund, *Perception, Mind and Personal Identity* (Lanham: University Press of America, 1994).

Sages and Scientists

Moses Maimonides quoted in David Burrell, *Knowing the Unknowable God: Ibn-Sina, Maimonides, Aquinas* (Notre Dame, IN: Notre Dame, 1986), 26, 39.
The Metaphysica of Avicenna transl. Parviz Morewedge (New York: Columbia University Press, 1973), 55.
Thomas Aquinas, *On Being and Essence.*
Madhvacharya quoted in B.N.K. Sharma, *A History of the Dvaita School of Vedanta and Its Literature* (New Delhi: Motilal Banarsidass, 2000), 159.
Thomas Aquinas quoted in Fritjof Capra, *The Tao of Physics* (Boston: Shambala, 1991), 287.
Stanley L. Jaki, *The Road of Science and the Ways to God* (Chicago: University of Chicago Press, 1978).
Stanley L. Jaki, *Science and Creation: From Eternal Cycles to an Oscillating Universe* (Edinburgh: Scottish Academic Press, 1986).
Stanley L. Jaki, *The Origin of Science and the Science of its Origin* (South Bend, Indiana: Regnery Gateway, 1978).

2. Two Warring Visions of the World

Theism

B.N.K. Sharma, *A History of the Dvaita School of Vedanta and Its Literature* (New Delhi:Motilal Banarsidass, 2000), 58-59.

3. The Three Foundations of Modern Science

W. Schmidt, *The Origin and Growth of Religion,* (London: Methuen, 1935)

The Three Foundations of Modern Science

Faith and Values in Science and Religion: A Discussion with Charles H. Townes (San Francisco: The Bhaktivedanta Institute, 1997).

Prophets of the Matrix

Copernicus, *On the Revolutions,* transl. Charles Glenn Wallis (Chicago: University of Chicago Press, 1952), 549.

Kepler quoted in Job Kozhamthadam, *The Discovery of Kepler's Laws* (Notre Dame, IN: University of Notre Dame, 1994), 13,19.

Galileo quoted in W.P. Carvin, *Creation and Scientific Explanation* (Edinburgh: Scottish Academic Press, 1988), 49.

Isaac Newton, *The Mathematical Principles of Natural Philosophy,* Book III, Andrew Motte, transl. (London: H.D. Symonds, 1803), 310ff.

W.D. Niven, ed., *The Scientific Papers of James Clerk Maxwell* (Cambridge: Cambridge University Press, 1890), Vol. 1, 759.

Lord Kelvin and J.J. Thompson cited in Henry F. Schaefer III, *Scientists and Their Gods* (Institute for Religious Research, 1999).

Albert Einstein, *Lettres a Maurice Solovine reproduits en facsimile et traduits en français* (Paris: Gauthier-Vilars, 1956), 102-3.

Albert Einstein, *Ideas and Opinions,* transl. Sonja Bargmann (New York: Dell Publishing Company, 1973), 49.

Albert Einstein, *Ideas and Opinions,* 255.

Einstein quoted in Timothy Ferris, *Coming of Age in the Milky Way* (New York: William Morrow, 1988), 177.

Max Planck, *Where is Science Going?* transl. With biographical note by James Murphy (New York: W.W. Norton, 1977), 168.

Max Planck quoted in Charles C. Gillespie ed. *Dictionary of Scientific Biography* (New York: Charles Scribner's Sons, 1975), 15.

Werner Heisenberg, *Across the Frontiers,* transl. Peter Heath (San Francisco: Harper and Row, 1974), 213.

Werner Heisenberg, *Physics and Beyond* (San Francisco: Harper and Row, 1971). Excerpted in Timothy Ferris ed. *The World Treasury of Physics, Astronomy and Mathematics* (New York: Little, Brown and Company, 1991), 826.

Max Born quoted in Frederick E. Trinklein, *The God of Science,* 64, 80

P.A.M. Dirac quoted in *Scientific American,* May 1963.

Einstein and God
> Max Jammer, *Einstein and Religion* (Princeton, NJ: Princeton University Press, 1999),
> 138, 123, 150, 149, 111.
> G.S. Viereck, *Glimpses of the Great* (New York: Macauley, 1930), 374.

The Curious Case of Charles Darwin
> Charles Darwin, *The Origin of Species by means of Natural Selection* (London: Penguin, 1968), 460.
> Charles Darwin cited in F. Darwin ed. *The Life and Letters of Charles Darwin* (New York: Appleton, 1898), I, 278.
> Loren Eisely, *Darwin's Century* (New York: Doubleday, 1958), 197.

4. The Four Masters of the Matrix

Avicenna
> *La Metaphysique du Shifa* transl. G.C. Anawati (Paris: Vrin, 1978), 8.3, 342:10-15.
> *The Healing,* Sixth Treatise, Chapter 1.
> *The Healing,* First Treatise, Chapter 6.
> *The Healing,* First Treatise, Chapter 7.

Moses Maimonides
> *Guide for the Perplexed,* Book I, 57.
> Moses Maimonides cited in Ronald H. Isaacs, *Every Person's Guide Jewish Philosophy and Philosophers* (Northvale, New Jersey: Jason Aronson Inc., 1999), 49.
> *Guide,* Book I, Chapter 52.
> *Guide,* Part II, chapter 28.
> *Guide,* Book I, Chapter 60.

Thomas Aquinas
> *On Being and Essence,* chapters 4-5.
> *Summa Theologica,* Part I, Question II, Third Article.
> *Summa Theologica,* Part I, Question XIII, Second Article.
> *Summa Theologica,* Part I, 76:1, 75:2..
> *Disputed Questions on Truth,* I, 3, 11, 12.

Madhvacharya
> B.N.K. Sharma, *Madhva's Teachings in His Own Words* (Mumbai: Bharatiya Vidya Bhavan, 1997), 132, 48, 133, 122.

5. An Eight-Fold Path to a Theory of Everything

John Barrow, *Theories of Everything: The Quest for Ultimate Explanation* (Oxford: Oxford University Press, 1991), 210

Russell Stannard quoted in Paul Davies, *The Mind of God: The Scientific Basis for a Rational World* (New York: Simon and Schuster, 1992), 166-7.

Mitchell Feigenbaum quoted in John Horgan, *The End of Science* (New York: Helix, 1996), 222.

The Guru2Geek Dialogues

I. Seeing is Believing

2. Seeing with the Brain

Data on the eye from standard sources including the *Science and Invention Encyclopedia* (Westport, CT: H.S. Stuttman, 1987) and *Evolution: The Triumph of an Idea.*

David Berlinski, "Has Darwin Met His Match?" *Commentary*, December 2002, 34.

Gerald Schroeder, *The Science of God* (New York: Broadway Books, 1997), 91-2, 104-107.

Stephen Jay Gould, *The Structure of Evolutionary Theory* (Cambridge: Harvard University Press, 2002), 1129.

Carl Zimmer, *Evolution: The Triumph of An Idea* (New York: Harper-Collins, 2001), 121 ff.

II. The Invention of Nature

Who Holds the Patent on Quantum Fields and the Genetic Code?

Max Jammer, *Einstein and Religion* (Princeton, NJ: Princeton University Press, 1999), 80.

The IQ of the Universe

John Maynard Smith, "Life at the Edge of Chaos?" *New York Review of Books*, March 2, 1995.

1. Space and Time

There's No Such Thing as Space

Lee Smolin, *Three Roads to Quantum Gravity* (New York: Basic Books, 2001), 18, 112-115.

Stanley L. Jaki, *Is There a Universe?* (Liverpool: Liverpool University Press, 1993), 99-106, 121.

Jean-Pierre Luminet, Glenn D. Starkmann and Jeffrey R. Weeks, "Is Space Finite?" *Scientific American*, April 1999, 90ff.

Matter in Motion from Newton to Einstein

Stanley L. Jaki, *The Absolute and the Relative* (Lanham, MD: The University Press of America, 1988), 1ff.

Gerald Holton quoted in Richard Panek, "And Then There Was Light." *Natural History*, November 2002, 51.

Gerald Schroeder, *The Hidden Face of God* (New York: The Free Press, 2001, 26.

Paul Davies ed. *The New Physics* (Cambridge: Cambridge University Press, 1989).

Time Out?

Rudolf Carnap, "Intellectual Autobiography" in P.A. Schilpp (ed.), *The Philosophy of Rudolf Carnap* (La Salle, IL: The Library of Living Philosophers, 1963), 37-8.

Stephen Hawking, *A Brief History of Time,* (New York: Bantam, 1988), 8,166.

Lee Smolin, *Three Roads to Quantum Gravity* (New York: Basic Books, 2001), 20, 138-9.

David Braine, *The Reality of Time and The Existence of God* (Oxford: Clarendon Press, 1988), 27ff., 50ff.

J.R. Lucas, "A Century of Time" in Jeremy Butterfield ed. *The Arguments of Time* (Oxford: Oxford University Press, 1999).

J.R. Lucas, "The Open Future" in Raymond Flood and Michael Lockwood, ed., *The Nature of Time* (Oxford: Basil Blackwell, 1986).

J.R. Lucas, *Space, Time and Causality* (Oxford: Oxford University Press, 1985); *The Future* (Oxford: Basil Blackwell, 1989);

J.R. Lucas, *A Treatise on Time and Space* (London: Methuen, 1973).

Ian Barbour, *The End of Time* (London: Weidenfeld and Nicholson, 1999).

Jeremy Butterfield, "The End of Time?" in *British Journal for the Philosophy of Science*, 2002.

Huw Price, *Time's Arrow and Archimedes' Point: New Directions for the Physics of Time* (Oxford: Oxford University Press, 1996).

Peter Coveney and Roger Highfield, *The Arrow of Time* (New York: Ballantine, 1990).

"A Matter of Time." *Scientific American* special issue, September 2002.

Hawking's Time-Less History

Stephen Hawking, *A Brief History of Time*, (New York: Bantam, 1988), 141, 174.

Gregory Benford, "Leaping the Abyss: Stephen Hawking on black holes, unified field theory and Marilyn Monroe." *Reason* 4.02, April 2002, 29.

Stephen Hawking, *The Universe in a Nutshell* (New York: Bantam, 2001), 41.

Gregory Benford, "Leaping the Abyss," 26.

Keith Ward, God, *Chance and Necessity* (Oxford: OneWorld, 1996), 41-3.

William Lane Craig, "What Place, Then, For a Creator?: Hawking on God and Creation." *British Journal for the Philosophy of Science*, 1990, 473-491.

Robin Le Poidevin, "Creation in a Closed Universe, Have Physicists Disproved the Existence of God." *Religious Studies*, March 1991, 39-48.

George F.R. Ellis in Roy Abraham Varghese, ed., *Great Thinkers on Great Questions* (Oxford: OneWorld, 1998), 176-7.

Richard Swinburne, *Is There a God?* (Oxford: Oxford University Press, 1997), 64.

Stephen Hawking, *Black Holes and Baby Universes* (New York: Bantam Books, 1993), 172.

2. The Quantum Dimension

The Standard Model

"The Dawn of Physics Beyond the Standard Model," by Gordon Kane, Scientific American, June 2003, 68ff.

What Quantum Physics Tells Us About The World

Abraham Pais, *Niels Bohr's Times, in Physics, Philosophy, and Polity* (Oxford: Oxford University Press, 1991).

David Deutsch, *The Fabric of Reality: The Science of Parallel Universes—And its Implications* (New York: Allen Lane, 1997).

John Bell, "Quantum Mechanics for Cosmologists" in Christopher Isham, Roger Penrose, Dennis Sciama, editors, *Quantum Gravity II* (Oxford: Clarendon Press, 1981), 611-637.

Jeremy Butterfield, "The End of Time?" in *British Journal for the Philosophy of Science,* 2002.

E.P. Wigner, "Reminiscences on Quantum Theory." Colloquium talk at Washington University, St. Louis, March 27, 1974.

Henry J. Folse, *The Philosophy of Niels Bohr: The Framework of Complementarity,* 1985.

John Honner, *The Description of Nature: Niels Bohr and the Philosophy of Quantum Physics,* 1987.

Dugald Murdoch, *Niels Bohr's Philosophy of Physics,* 1987.

E.T. Jaynes, "Probability in Quantum Theory" in W.H. Zurek, ed., *Complexity, Entropy and the Physics of Information* (Reading, MA: Addison-Wesley Publishing Co., 1990).

C.A. Fuchs and A. Peres, "Quantum Theory Needs No Interpretation." *Physics Today* 53, 2000.

C.A. Fuchs, "Quantum Mechanics as Quantum Information (and only a little more)." (paper) "The Structure of Quantum Information." (paper)

C.A. Fuchs, "Notes on a Paulian Idea: Foundational, Historical, Anecdotal and Forward-Looking Thoughts on the Quantum."

Peter Hodgson, "God's Action in the World: The Relevance of Quantum Mechanics." *Zygon: Journal of Religion and Science,* September 2000.

Grete Hermann, *The Harvard Review of Philosophy,* Spring 2000.

Gordon Kane, "The Dawn of Physics Beyond the Standard Model," Scientific American, June 2003, 68ff.

Julian Schwinger, Relativistic Quantum Field Theory, Nobel Lecture, December 11, 1965.

Robert Spitzer, "Proofs for the Existence of God—Part I: A Metaphysical Argument." *International Philosophical Quarterly,* June 2001, 162-186.

Robert Spitzer, "Proofs for the Existence of God—Part II." *International Philosophical Quarterly,* September 2001, 305-331.

Adolf Grunbaum, "The Pseudo-Problem of Creation." *Philosophy of Science,* September 1989, 383-5.

જ

3. The Big Bang and Before

P. James E. Peebles, "Making Sense of Modern Cosmology." *Scientific American*, January 2001.

The Basis of Big Bang Theory

P.J.E. Peebles, *Principles of Physical Cosmology* (Princeton, NJ: Princeton University Press, 1993).

John D. Barrow, *The Origin of the Universe* (New York: Basic Books, 1994).

Leon M. Lederman and David N. Schramm, *From Quarks to the Cosmos* (New York: Scientific American Library, 1995).

"First Year Wilkinson Microwave Anisotropy Probe (WMAP) Observations: Determination of Cosmological Parameters," D.N. Spergel, et al. (2003).

From Whimper to Bang and Back

Alan Guth, *The Inflationary Universe: The Quest for a New Theory of Cosmic Origins* (New York: Perseus Press, 1997).

Mario Livio and Allan Sandage, *Accelerating Universe: Infinite Expansion, the Cosmological Constant, and the Beauty of the Cosmos* (New York: John Wiley and Sons, 2000).

Alan Lightman and Roberta Brawer ed. *Origins: The Lives and Worlds of Modern Cosmologists* (Cambridge, MA: Harvard University Press, 1990).

J.J. Halliwell, *Quantum Cosmology* (Cambridge: Cambridge University Press, 1991).

Andrei Linde, "The Self-Reproducing Inflationary Universe." *Scientific American*, March 1998.

"First Year Wilkinson Microwave Anisotropy Probe (WMAP) Observations: Implications for Inflation," H.V. Peiris, et al. (2003).

Martin Rees cited in *New Scientist*, February 1, 2003.

Alan Guth, Andrei Linde, Alexander Vilenkin, Saul Permutter, Martin Rees, J. Richard Gott cited in "What Came Before Creation?" *U.S. News & World Report* special edition on Mysteries of Science, 2002.

Willem B. Drees, *Beyond the Big Bang* (La Salle, Illinois: Open Court, 1990), 22-24.

James Trefil, *101 Things You Don't Know About Science* (New York: Mariner, 1996), 114.

The Anthropic Principle

Martin Rees, *Just Six Numbers* (New York: Basic Books, 2000).

John C. Gribbin and Martin Rees, *Cosmic Coincidences* (New York: Bantam, 1989).

Richard Swinburne, *Is There a God?* (Oxford: Oxford University Press, 1997), 67-8.

John Barrow cited in *New Scientist*, September 7, 2002, 33.

Paul Davies, *The Mind of God: The Scientific Basis for a Rational World* (New York: Simon and Schuster, 1992), 220.

Paul Davies, *The Cosmic Blueprint* (New York: Simon and Schuster, 1988).

Richard Swinburne, *The Coherence of Theism* (Oxford: Clarendon Press, 1977).

Richard Swinburne, *The Existence of God* (Oxford: Clarendon Press, 1979).

John Leslie, *Infinite Minds, A Philosophical Cosmology* (Oxford: Oxford University Press, 2001).

4. DNA and the "Facts" of Life

Family Tree

Carl Zimmer, *Evolution: The Triumph of An Idea* (New York: HarperCollins, 2001), 66ff.

Richard Fortey, *Life: A Natural History of the First Four Billion Years of Life on Earth* (New York: Alfred A. Knopf, 1998).

Life in the Universe, Scientific American special issue, October 1994.

Richard Swinburne, *Is There a God?* (Oxford: Oxford University Press, 1997), 62-3.

The Nature and Origin of Life

Josef Seifert, *What is Life?* (Amsterdam-Atlanta, GA: Rodopi, 1997), 34-61

E.J. Ambrose, *The Mirror of Creation* (Edinburgh: Scottish Academic Press, 1990), 84ff.

Ian Stewart, *Life's Other Secret: The New Mathematics of the Living World* (New York: Wiley, 1998).

David Berlinksi, *The Advent of the Algorithm* (New York: Harcourt, 2000), 286-305.

Gerald Schroeder, *The Hidden Face of God: How Science Reveals the Ultimate Truth* (New York: The Free Press, 2001), 189 ff.

Tony Trewavas quoted in *New Scientist*, July 27, 2002.

Stuart Kaufmann quoted in *New Scientist*, June 13, 1998.

Francis Crick cited in John Maddox, *What Remains to be Discovered* (New York: Touchstone, 1998), 131, 142-3.

Stanley Miller quoted in *Earth*, February 1998.

Gunter Wachstershauser and Jeffrey Bada quoted in *Earth*, February 1998.

RNA-based origin of life studies described in *New Scientist* May 17, 2001.

Lyn Margulis quoted in *New Scientist*, June 13, 1998.

Eckard Wimmer quoted in *Business Week*, July 22, 2002.

Stuart Kaufmann, *The Origins of Order: Self-Organization and Selection in Evolution* (New York: Oxford University Press, 1993).

Stuart Kaufmann, *At Home in the Universe* (New York: Oxford University Press, 1995).

Stuart Kaufmann cited in *New Scientist* July 6, 1996.

Interview with Stuart Kaufmann in John Horgan, *The End of Science* (New York: Helix, 1996), 132-7.

Stuart Kaufmann's origin of life studies cited in John Maddox, *What Remains to be Discovered* (New York: Touchstone, 1998), 141-2

Paul Davies, *The Fifth Miracle: The Search for the Origin and Meaning of Life* (New York: Touchstone, 1999), 139-141.

John Maddox, *What Remains to be Discovered* (New York: Touchstone, 1998), 369-370.

Werner Arber, "The Existence of a Creator Represents a Satisfactory Solution" in Henry Margenau and Roy Abraham Varghese ed. *Cosmos, Bios, Theos* (La Salle: Open Court, 1992), 142.

5. The Biological Blueprint

Evolutionary Mechanisms and Purposive Structures

Keith Stewart Thomson, "The Meanings of Evolution." *American Scientist*, September/October 1982.

J.J.C. Smart and John Haldane, *Atheism and Theism* (Great Debates in Philosophy) (Oxford: Blackwell Publishers, 2003), 224.

Peter Geach, *Providence and Evil* (Cambridge: Cambridge University Press, 1977).

John Maddox, *What Remains to be Discovered* (New York: Touchstone, 1998), 252.

Michael Ruse, *Darwin and Design: Does Evolution Have a Purpose?* (Cambridge, MA: Harvard University Press, 2003).

Peter McLaughlin, *What Functions Explain* (Cambridge: Cambridge University Press, 2001).

Kenneth Gallagher, " 'Natural Selection' A Tautology?" *International Philosophical Quarterly*, March 1989.

Robert Williams, J.F. DaSilva and Harold Morowitz cited in *New Scientist*, January 2003.

Werner Arber, "Traditional Wisdom and Recently Acquired Knowledge in Biological Evolution," SSQ Conference, UNESCO, Paris, April 2002.

Gerald Schroeder, *The Hidden Face of God* (New York: The Free Press, 2001, 120-1

Gerald Schroeder, *The Science of God* (New York: Broadway Books, 1997), 29-30, 113

Peter Brown, "Engines of Evolution." *Natural History*, February 2003, 6.

John Maddox (on genes), *What Remains to be Discovered* (New York: Touchstone, 1998), 153-4, 165-6, 175-8, 369-370

Niles Edredge, *The Pattern of Evolution* (New York: W.H. Freeman, 1998).

Richard Dawkins, "God's Utility Function." *Scientific American*, November 1995, 80ff.

Stephen Jay Gould, *The Structure of Evolutionary Theory*, (Cambridge: Harvard University Press, 2002), 613ff.

Hominids, Gorillas and Chimpanzees

Jonathan Marks, *What It Means to Be 98% Chimpanzee: Apes, People and Their Genes* (Berkeley: University of California Press, 2002), 1ff, 161.

Interview with Jonathan Marks cited in Robert S. Boyd, "Researchers finding more ways chimps resemble humans." Knight Ridder, September 25, 2002.

Elaine Morgan, *The Aquatic Ape: A Theory of Human Evolution* (London: Souvenir Press, 1989), 17-8.

The Existence of the Mind

David H. Lund, *Perception, Mind and Personal Identity* (Lanham: University Press of America, 1994).

Richard Swinburne, "The Origin of Consciousness" in Clifford N. Matthews and Roy Abraham Varghese ed. *Cosmic Beginnings and Human Ends* (Chicago: Open Court, 1995), 358.

Sir John Eccles, "A Divine Design: Some Questions on Origins" in Henry Margenau and Roy Abraham Varghese ed. *Cosmos, Bios, Theos* (La Salle: Open Court, 1992), 164

John Maddox, *What Remains to be Discovered* (New York: Touchstone, 1998), 370.

Steven Pinker, *How the Mind Works* (New York: W.W. Norton, 1997), 131-148.

J.J.C. Smart and John Haldane, *Atheism and Theism* (Great Debates in Philosophy) (Oxford: Blackwell Publishers, 2003), 228 ff.

Russell Pannier and T.D. Sullivan in Roy Abraham Varghese ed. *Great Thinkers on Great Questions* (Oxford: OneWorld, 1998), 137.

Heini K.P. Hediger, "The Clever Hans Phenomenon from an Animal Psychologist's Point of View" in *The Clever Hans Phenomenon*, 5, 9.

Ian Tattersall, *The Monkey in the Mirror: Essays on the Science of What Makes Us Human, Scientific American* Dec 2001, 58ff.

Herbert McCabe, "Sense and Sensibility." *International Philosophical Quarterly*, December 2001.

Richard Dawkins, *Unweaving the Rainbow* (Boston, MA: Houghton Mifflin Company, 1998), 286 ff.

Stephen R.L. Clark, "The Evolution of Language: Truth and Lies." *Philosophy*, July 2000, 408-9

John Leslie, *Infinite Minds, A Philosophical Cosmology* (Oxford: Oxford University Press, 2001).

J.R. Lucas, *The Freedom of the Will* (Oxford: Clarendon Press, 1976).

Peter Geach, *Providence and Evil* (Cambridge: Cambridge University Press, 1977).

John Eccles and Daniel N. Robinson, *The Wonder of Being Human: Our Brain and Our Mind* (Boston, MA: New Science Library, 1985), 17.

K.R. Popper and J.C. Eccles, *The Self and Its Brain* (Heidelberg: Springer Verlag International, 1977).

C.D. Broad, *The Mind and Its Place in Nature* (London: Kegan Paul, 1925), 623.

David J. Chalmers, *The Conscious Mind: In Search of a Fundamental Theory*, Oxford University Press, November 1997.

Colin McGinn, *The Mysterious Flame: Conscious Minds in a Material World*, Basic Books, April 25, 2000.

☙

6. Interacting with Infinite Intelligence

Karl Popper cited in John Eccles and Daniel N. Robinson, *The Wonder of Being Human: Our Brain and Our Mind* (Boston, MA: New Science Library, 1985), 18.

K.R. Popper and J.C. Eccles, *The Self and Its Brain* (Heidelberg: Springer Verlag International, 1977).

J.J.C. Smart and John Haldane, *Atheism and Theism* (Great Debates in Philosophy) (Oxford: Blackwell Publishers, 2003), 230.

Ellen Ullman, "Programming the Post-Human." *Harper's Magazine*, October 2002, 65.

7. The Laws of Nature

Albert Einstein, *Ideas and Opinions,* transl. Sonja Bargmann (New York: Dell Publishing Company, 1973), 255.

Einstein on the "thoughts" of God quoted in Timothy Ferris, *Coming of Age in the Milky Way* (New York: William Morrow, 1988), 177.

Stephen Hawking, *A Brief History of Time,* (New York: Bantam, 1988), 175.

Paul Davies, *The Mind of God: The Scientific Basis for a Rational World* (New York: Simon and Schuster, 1992), 31.

Martin Rees, "Exploring Our Universe and Others" in *The Frontiers of Space* (New York: Scientific American, 2000), 87.

Paul Davies on John Archibald Wheeler, *The Mind of God,* 225.

Steven Weinberg, *New York Review of Books,* May 31, 2001.

Paul Davies, *The Mind of God,* 93ff.

Roger Penrose, *The Emperor's New Mind* (Oxford: Oxford University Press, 1989), 143

Gödel's ontological argument cited in John W. Dawson, Jr., "Gödel and the Limits of Logic." *Scientific American,* June 1999, 81.

Paul Davies, "What Happened Before the Big Bang?" in Russell Stannard ed. *God for the 21ˢᵗ Century* (Philadelphia: Templeton Foundation Press, 2000), 12.

Richard Swinburne in Roy Abraham Varghese ed. *Great Thinkers on Great Questions* (Oxford: OneWorld, 1998), 117-120.

Keith Ward in Roy Abraham Varghese ed. *Great Thinkers on Great Questions* (Oxford: OneWorld, 1998), 184-190.

Heinz Pagels, *The Cosmic Code* (New York: Bantam, 1984), 84 ff.

Stanley L. Jaki, *Is There a Universe?* (Liverpool: Liverpool University Press, 1993), 100, 122

Herbert McCabe, "Sense and Sensibility." *International Philosophical Quarterly*, December 2001, 815.

III. The Mind of God

1. Can Our Minds Tell the Truth?
T. Theocharis and M. Psimopoulos, "Where Science Has Gone Wrong." *Nature*, October 15, 1987, 595-8.
Illtyd Trethowan, *Absolute Value* (London: George Allen & Unwin Ltd, 1970), 5
T. Theocharis and M. Psimopoulos, "Where Science Has Gone Wrong." *Nature*, 598.

2. Monisms of the East—Past and Present

Hinduism, Buddhism, Taoism
Radhakrisnan cited in B.N.K. Sharma, *A History of the Dvaita School of Vedanta and Its Literature* (New Delhi: Motilal Banarsidass, 2000), 22.
Sharma on theism in the *Upanishads, A History of the Dvaita School of Vedanta and Its Literature,* 24, 32-4.
Geoffrey Parrinder, *Mysticism in the World's Religions* (Oxford: One-World, 1995), 94.
Sanghamitra Sharma, *Legacy of the Buddha: The Universal Power of Buddhism* (Mumbai: Eeshwar, 2001)
Eknath Eshwaran, transl., *The Dhammapada* (New Delhi: Penguin Books, 1986).
Radhakrishnan and Dasgupta cited in B.N.K. Sharma, *A History of the Dvaita School of Vedanta and Its Literature,* 63.

Madhva's Critique of Monism
B.N.K. Sharma, *A History of the Dvaita School of Vedanta and Its Literature* (New Delhi: Motilal Banarsidass, 2000), 121, 94, 123, 124.

The Physics and Philosophy of the Tao
Fritjof Capra, *The Tao of Physics* (Boston: Shambala, 1991), 286.

Gary Zukav, *The Dancing Wu Li Masters: An Overview of the New Physics* (New York: Bantam, 1980).

Tony Rothman and George Sudarshan, *Doubt and Certainty* (New York: Helix, 2001), 47.

Frederick Copleston, *Religion and the One* (New York: Crossroads, 2003), 66.

Eric R. Scerri, "Eastern Mysticism and the Alleged Parallels with Physics." *American Journal of Physics*, August 1989, Volume 57, Number 8, 687ff

J. Bell, *Speakable and Unspeakable in Quantum Mechanics* (Cambridge: Cambridge University Press, 1987), 170.

Schrödinger cited in Eric R. Scerri, "Eastern Mysticism and the Alleged Parallels with Physics," 689.

Fritjof Capra, *The Tao of Physics,* 286

Sal P. Restivo, "Parallels and Paradoxes in Modern Physics and Eastern Mysticism." *Social Studies of Science*, May 1978, 143ff.

3. From Cosmos to Theos

Richard Swinburne, "The Limits of Explanation" in *Explanation and its Limits* (Cambridge: Cambridge University Press, 1990).

William Stoeger, "The Origin of the Universe in Science and Religion" in Henry Margenau and Roy Abraham Varghese ed. *Cosmos, Bios, Theos* (La Salle: Open Court, 1992), 263

Hugo Meynell, *The Intelligible Universe* (Totowa, New Jersey: Barnes and Noble, 1982), 104-5.

Bertrand Russell and F.C. Copleston, "A Discussion on the Existence of God" in *The Existence of God* (New York: Macmillan, 1964), 175

Jean Paul Sartre, *Being and Nothingness* (London: Methuen, 1957), 539-540.

J.J.C. Smart and John Haldane, *Atheism and Theism* (Great Debates in Philosophy) (Oxford: Blackwell Publishers, 2003), 228 ff.

4. Infinite Perfection

H.D. Lewis, *Philosophy of Religion* (London: The English Universities Press, 1965), 145.

5. Before Time and Outside Space

Illtyd Trethowan, *The Absolute and the Atonement* (London: George Allen and Unwin, 1971), 156.

Brian Leftow, *Time and Eternity* (Ithaca, NY: Cornell University Press, 1991).

Brian Leftow in Roy Abraham Varghese ed. *Great Thinkers on Great Questions* (Oxford: OneWorld, 1998), 230-2.

6. The Problem of Evil

Sandra Menssen and T.D. Sullivan in Roy Abraham Varghese ed. *Great Thinkers on Great Questions* (Oxford: OneWorld, 1998), 202-4.

7. The Denial of the Divine

Paul C. Vitz, *Faith of the Fatherless: The Psychology of Atheism* (Dallas: The Spence Publishing House, 1999).

A Hundred Wonders of the World

51. Insect Flight

Michael Dickinson, "Solving the Mystery of Insect Flight," *Scientific American*, June 2001, 49ff.

64. Seeing–2

Reference to computer simulation of retinal nerve cells in John Stevens, *Byte*, April 1985.

INDEX

Acknowledgements

I wish to thank all those who have provided invaluable information, advice and assistance in the creation of this work, most especially the following:

1. Swami Agnivesh
2. Werner Arber
3. Sir Alfred Ayer
4. George Brody
5. Will Burkett
6. F.F. Centore
7. Sir John Eccles
8. Kazimierz Dadak
9. Antony Flew
10. Mike Grimshaw
11. Raj Mohan Gandhi
12. J.J. Haldane
13. Daniel K. Hennessy
14. John Lenihan
15. J.R. Lucas
16. Robert Jastrow
17. James R. Jones
18. William LaMothe
19. Henry Margenau
20. Paul Massell
21. Charles Phipps
22. Alvin Plantinga
23. Leon Pugh
24. Daniel R. Robinson
25. Gerald Schroeder
26. Josef Seifert
27. B.N.K. Sharma
28. Mark Shea
29. Richard Swinburne
30. Jodie and Dottie Thompson
31. Charles Townes
32. Anila Varghese
33. Rachel, Mary and Michael Varghese
34. M.M. Varghese, M. Abraham Varghese and Leela Abraham
35. W. Marvin Watson

Visit us on the Web at

http://www.thewonderoftheworld.com/

SPECIAL ONLINE FEATURES:

♦ Message from the author

♦ Interview with the author

♦ Glossary of terms found in the book

♦ Biographies of scientists, philosophers
  and theologians cited in the book

♦ Multimedia presentation of the 100
  Wonders of the World

♦ and more!